GEOLOGICAL EXCURSIONS
AROUND MIRI, SARAWAK

ISBN
978-1-4828-5486-2 (sc)
978-1-4828-5487-9 (e)

Print information available on the last page.

To order additional copies of this book, contact
Toll Free 800 101 2657 (Singapore)
Toll Free 1 800 81 7340 (Malaysia)
www.partridgepublishing.com/singapore
orders.singapore@partridgepublishing.com

06/09/2016

PARTRIDGE

Cover page photo: The cliffs of Tanjung Lobang, near Miri, with seaward dipping sandstones of the Miri Formation (photo by Mario Wannier)

GEOLOGICAL EXCURSIONS AROUND MIRI, SARAWAK

1910 – 2010: CELEBRATING THE 100th ANNIVERSARY OF THE DISCOVERY OF THE MIRI OIL FIELD

Mario Wannier
Philip Lesslar
Charlie Lee
Han Raven
Rasoul Sorkhabi
Abdullah Ibrahim

About the Authors

Mario Wannier holds a PhD in Earth Sciences from Basle University. He works for Shell International Exploration and Production, and has over 30 years experience in Exploration; he has been involved in a variety of projects in Europe, Africa, South East Asia and the Arctic. He worked and lived in Miri for over 5 years. Mario is currently Team Leader for Exploration New Ventures in Russia and Ukraine

Charlie Lee is a geologist with Shell (since 1996), and currently works as Staff Seismic Interpreter, exploring the poorly imaged sub-salt play in the Gulf of Mexico. A native of Malaysia, he has studied and lived in Australia, Malaysia and the USA. The main focus of his work in Shell is deepwater exploration evaluation and prospect maturation, with projects spanning across Asia-Pacific, North Africa and the Americas. A keen fossil hunter, he has a large collection of Miocene marine fossils from Borneo, in particular fossil crabs and mollusks.

Han Raven holds a PhD from Leiden University. Since 1984 he works as a geologist for Shell International Exploration and Production, currently as Global Consultant Front End Project Management. Since his childhood he has studied recent and Cenozoic mollusks, especially the relationship between mollusk faunas and depositional environments. He is a research associate of the Netherlands Centre for Biodiversity Naturalis in Leiden, The Netherlands. From 1992 to 1997 he worked and lived in Miri. In his spare time he studied Miocene to recent shallow-water faunas of Sarawak, Brunei and Sabah. Based on his sample collections, he is involved in several studies, including sea-level changes during the Holocene and related changes in depositional environments and faunas. He has published various papers on specific mollusks groups and several new papers are in preparation. More information can be found at http://science.naturalis.nl/research/people/cv/raven

Dr. Rasoul Sorkhabi is a research professor at the University of Utah's Energy & Geoscience Institute, Salt Lake City (since 2003). A native of Iran, he has studied and lived in India, Japan, and the USA. Prior to joining the University of Utah, he worked for Japan National Oil Corporation and Arizona State University. His research fields include petroleum geology, structural geology, geochronology, and the history of geology. He has conducted field works in various parts of the world including Sarawak and Sabah. He has co-edited Tectonophysics Special Volume 451 (2008), AAPG Memoir 85 (2005), and GSA Special Paper 329 (1999). He is a contributing editor for the magazines Earth and GeoExpro.

Philip Lesslar is a Miri local and has worked for Shell Malaysia Exploration & Production for 36 years. He started his Shell career in micropaleontology, working on the Tertiary foraminifera of NW Borneo. After 13 years in this field, he made a career switch to geological information management. In this capacity, he has worked and lived in Houston, Texas, and in Rijswijk, The Netherlands and was responsible for the development and deployment of Global EP Data Standards and Data Quality Metrics.
He is a member of IAIDQ (International Association of Information & Data Quality) and has served on the program committees for data quality conferences.

Abdullah Ibrahim currently works as a Deepwater Explorationist for Brunei Shell Petroleum. The subject of his MSc thesis delivered 1998 was the depositional evaluation of the Liang Formation in Brunei Darussalam. Since then he has periodically revisited the quarry locations and basked in the sights of significant sandy deposits that make up the Liang formation. His current mission in life since 1998, however, has largely been focused in finding similarly magnificent sandy deposits in deepwater environments, particularly in Offshore Brunei.

Some Notes

Authorship

Being a product of team work, a note on the authorship of this publication is apt here. Chapters 1, 2 and 3 were largely written by Mario Wannier, with contributions from Han Raven (chapters 1.3.4, 1.3.5, 2.3.4 and 3.11), Rasoul Sorkhabi (chapters 1.2, 2.2 and 3.4.5), and Abdullah Ibahim (chapter 3.14). Chapter 4 was largely written by Charlie Lee and Han Raven, with contributions from Mario Wannier (chapters 4.1.3, 4.1.4 and 4.9). Philip Lesslar and Charlie Lee wrote chapter 5. Chapters 6 and 7 were largely written by Mario Wannier with contributions from Rasoul Sorkhabi. The illustrations have been provided by the authors. Rasoul Sorkhabi critically read and revised the whole manuscript. Final editing was carried out jointly by Mario Wannier and Rasoul Sorkhabi. Philip Lesslar handled the final coordination, quality control and management of this project to bring the book to completion.

Acknowledgements

We gratefully acknowledge contributions by Musa Musbah (Miri) on Nannoplankton dating, Jan Wilschut (The Hague) on the Brunei tektites, and Bert Hoeksema (Netherlands Centre for Biodiversity Naturalis, Leiden, The Netherlands) on the identification of coral samples. Rod Nourse and Maarten Wiemer critically reviewed the initial draft and offered valuable comments to improve its content.

Abbreviations & Terms

Abbreviations for "million years ago" (Ma) and "before present" (BP) have been used throughout the text. Other abbreviations include "barrels of oil" (bbls), "barrels of oil per year" (bbls/y), and "million barrels" (MMB); for locations of outcrops along roadside, "left" (L) and "right" (R) abbreviations have been used.

We have used "Shell" as a generic company name for a number of subsidiaries historically active in Borneo, such as the Anglo-Saxon Petroleum Company, the Asiatic Petroleum Company, Sarawak Oilfields Limited (SOL), Bataafsche Petroleum Maatschappij (BPM), British Malayan Petroleum Company Limited, etc

For ease of communication, we use "Tertiary" instead of the recommended term of "Cenozoic" (which includes both "Tertiary" and "Quaternary").

We have tried to be explicit about scientific concepts, but the use of technical jargon is sometimes inevitable. For this reason, we have added a geological glossary at the end of this publication.

Disclaimer

This publication has been prepared as a guide to the geological sites in and around Miri. Travelling to and exploring the outcrops described here carries inherent risks. Before visiting any of the outcrops described in this publication, proper planning, working knowledge of the area and common sense are required. Since the time of writing, access to some outcrops may have been rendered more difficult and hazardous, and new unforeseen dangers and safety issues may have developed. The authors disclaim any liability in negligence or otherwise for any loss, injury or damage that may occur as a result of reliance on any information contained in this publication.

And a Request ...

While the authors have made their best to provide accurate information and to produce a useful publication, there may be errors in this report. Please let us know if any corrections need to be made or some new information should be added. We will incorporate such modifications and updates in the second edition of the guidebook. Thank you!

Table of Contents

Chapter 1

Miri and the Birth of the Oil Industry in Malaysia

1.1 Remembering Miri Well#1

Anniversaries are good occasions to remember the significant achievements of the past and to cherish these memorable events for ours and generations to come. However, some historical events have had such an important impact on a period, place or society that their significance goes beyond anniversaries. The story of the Miri oil field is such a case (Sarawak Shell Berhard, 1990; Sorkhabi, 2010). The discovery of this oil field on 22nd December 1910 created enormous wealth not only for those involved with its discovery, but also for the state and citizens of Sarawak. The prize of the field was such that it has been called, in a typical colonialist style, *"the most prolific oil field in the whole of the British Empire"* (Hose, 1929), and it became one of the first targets of the Japanese occupation of Borneo in December 1941. The activities associated with the development and production of the oil field in Miri caused parallel development of a vibrant local economy and general improvements in the living standards of the Sarawak population (Jackson, 1968). The first doctor was on duty in 1919, and the first Miri hospital was built in 1920, offering medical services to oilfield workers as well as to the people of the Fourth Division of Sarawak (Harper, 1975).

Miri was a small settlement in the nineteenth century. While the pre-war growth of Miri was considerable, largely due to the Miri oil field, the economic and political expansion the city has witnessed in recent decades is truly remarkable. Today, Miri is one of the most modern and rapidly growing cities in Borneo. Its new international airport is used by about 2 million passengers a year! Much of this new wealth and prestige is due to the successful presence of the oil industry, now expanding beyond the Miri oil field and into the offshore.

The natural riches of the region go far beyond petroleum. Several National Parks and forest sites, which are accessed from Miri, attract an ever-growing flock of tourists, as well as scientists and researchers. Today, Miri offers all levels of education; it houses comfortable hotels and shopping malls; it has many museums, including a Petroleum Museum (Figures 1.1a and 1.1b), as well as the first protected geological outcrop (the Miri

Figure 1.1a: The Petroleum Museum, on top of Canada Hill, with the reconstructed derrick of Miri#1

2

Airport Road Outcrop) in the country.

As we enjoy these developments and recreations that Miri offers, we can also learn about the geologic story of rocks that make up Miri's landscape, and its celebrated oil field and the pioneers who discovered and developed it.

In 1960, Shell geologist P. Liechti and co-workers published a landmark report on the geology of Sarawak, Brunei and the western part of Borneo (Liechti et al., 1960). In his foreword to the report, S.H. Shaw, Director of Overseas Geological Surveys in London, remarked: "This publication gives the first detailed account of the geology of more than 58,000 square miles of Sarawak, Brunei, and western North Borneo. Appropriately its issue marks 50 years of scientific exploration in this region by the Royal Dutch Shell Group of oil companies, whose activities started in 1910 with their discovery of the Miri oilfield". Much geologic work has been done in Miri since then, and the place still offers a large number of valuable sites for geologic education and field excursions in Borneo.

This publication attempts to introduce the salient features of the geology of the Miri region to the general public, geology students and geologists who are visiting the city to look at outcrops. This guidebook, published on the occasion of the centenary of the discovery well Miri#1, hopefully provides a geological as well as a historical perspective on the way science and business complement each other. It is the further hope of the authors that this guidebook, used in conjuction with field trips in Miri, will attract young talents to the fascinating world of geology.

Figure 1.1b: The silhouette of Miri#1 as it appeared in the 1990s

1.2 Miri in History

Sarawak is located on the northwest coast of Borneo, the world's third largest island (Figure 1.2.1). It is rich in natural resources; indeed the word "serawak" is a Malay word for the mineral antimony. Paleolithic cavemen and hunter-gatherer tribes first settled in Sarawak; the oldest evidence includes a *Homo sapiens* skull from the Niah Caves near Miri, estimated to be 40,000 years old. About 4,500 years ago migrating waves of Austronesians, the ancestors of the present Dayak peoples, came to this region. Chinese and Malay traders visited Sarawak as early as 900 A.D., bringing Buddhist, Hindu and Muslim traditions to Borneo. The Europeans came first in the 16th century. In 1512, Antonio Pigafetta, an Italian companion and chronicler of Ferdinand Magellan, wrote an account of the region, of what he called "Cerava." In the 1820s, Dutch colonists began to exert their influence in Kalimantan, the southern part of Borneo.

The modern history of Sarawak began in 1839 when James Brooke arrived in Sarawak. Brooke was born in 1803 in India and was an army officer of the British East India Company. Upon his father's death, James quit his job and used his inheritance to purchase a schooner, the *Royalist*, and sailed for Sarawak. There, he met with Rajah Muda Hashim, who was governing the region on behalf of his nephew, Sultan Omar Ali Saifuddin II of Brunei. As Brooke's party left Sarawak for Singapore, Dayak warriors unsuccessfully attacked his ship. In 1840, Rajah Muda Hashim requested Brooke's help to defeat a Dayak revolt in Kuching, promising in return to give him Kuching and environs (later called the First Division of Sarawak). Following this success, Brooke was appointed the first White Rajah of Sarawak in 1841, in exchange

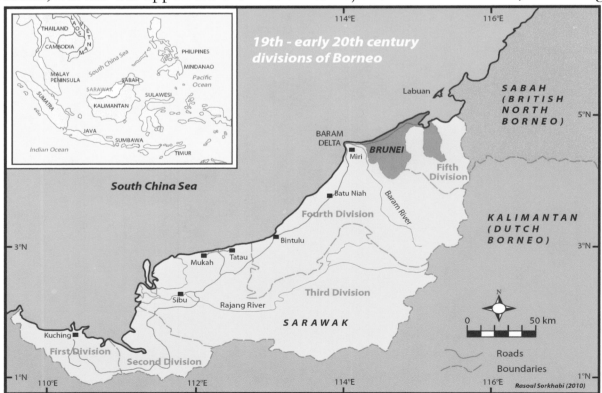

Figure 1.2.1: Map of Sarawak showing the 19th to early 20th century political divisions

for a small annual payment to the Sultan of Brunei. Initially the British were not supportive of Brooke, but after Brooke began to eradicate piracy in the South China Sea, and following the Sultan of Brunei's decision, in 1846, to cede the nearby island of Labuan to the British Royal Navy as a base for fighting the local pirates, relationships between Brooke and Britain were cemented. Historical accounts of the White Rajahs of Sarawak have been given in Runciman (1960), Pybus (1960), Reece (1982), and Payne (1997).

The first geological survey of Sarawak was conducted by Hiram Williams in 1845. Sarawak was also part of Alfred Russel Wallace's natural history exploration from 1854-1862, which is reported in his famous book *The Malay Archipelago* (1869). Wallace refers to coal mines in Sarawak: "These [coal mines] puzzle the natives exceedingly, as they cannot understand the extensive and costly preparations for working coal, and cannot believe it is to be used only as fuel when wood is so abundant and so easily obtained". The Italian biologist Odoardo Beccari also explored Sarawak from 1865-68, as documented in his *Nelle Foreste di Borneo* (1902; English translation: *Wanderings in the Great Forests of Borneo*, 1904).

The Brooke administration was in business engagements with companies in Scotland and Singapore. To better facilitate trade between Sarawak and Britain, the Borneo Company Limited was founded in London in 1856. In the following decades, surveyors working for this company carried out extensive mapping of Sarawak for its mineral and natural resources (Henry Longhurst details this history in his 1957 book *The Borneo Story: The First Hundred Years of the Borneo Company*).

In 1864, Britain recognized Sarawak as an independent state. In 1868, James Brooke was succeeded by his nephew Charles Johnson (later renamed Brooke), who ruled Sarawak until 1917 and, over time, extended his domain to the present boundaries of Sarawak. In 1888, the White Rajah obtained British protection for Sarawak. During Sir Charles Brooke's period, surveys and exploration in Sarawak increased. Dr. T. Posewitz published *Borneo: Its Geology and Mineral Resources* (London, 1892), and H. Ling Roth and H. Brooke Low produced a comprehensive description of *The Natives of Sarawak and British North Borneo* (London, 1896).

Like in the other oil regions of the world, oil seeps provided the first motivation for drilling in Sarawak. Indeed, local inhabitants had extracted oil from hand-dug wells for centuries. The 11th century *Song Hui Yao*, a historical compilation of the Song Dynasty of China, mentions imports of camphor and petroleum from Borneo. In 1882, Claude Champion de Crespigny, Resident (chief officer) of the Fourth Division of Sarawak (Baram), listed in his report to the Brooke Government 18 hand-dug oil wells in the Miri area. Perhaps foreseeing the value of what the locals called "minyak tanah" or "earth oil," De Crespigny recommended in his 1884 journal that "the oil district near the mouth of the Miri River should be thoroughly searched and reported on." "Earth oil" was used by local people for lighting lamps, waterproofing boats, and medicinal purposes. But De Crespigny's words fell on deaf ears in the Brooke government. However, his successor Charles Hose, a renowned naturalist and officer at the service of Sir Charles Brooke's government, pursued the Miri oil idea and his efforts as well as the entry of Shell changed the history of this region for ever.

1.3 Remembering the Pioneers

1.3.1 Charles Hose (1863-1929)

Colonial administrator, collector, ethnographer and naturalist, Charles Hose passed into posterity with glory, having one of Sarawak's mountain ranges named after him. The 21-year old Charles Hose arrived in Sarawak in 1884 and worked initially as a cadet under Rajah Charles Brooke (Figure 1.3.1a); he remained in service in Sarawak's 3rd Division, based for 20 years in Marudi, and 3 years in Sibu. In 1905, aged 42, he went into retirement in England.

As part of his duties, Charles Hose travelled all over northern Sarawak and became familiar with the local populations and their cultures, as well as with the overpowering natural environment he was confronted with daily. He published abundantly on his observations and was recognized as a leading expert on Sarawak natural history.

Figure 1.3.1a: Charles Hose as a cadet, 1884 (from Hose, C., 1927)

While on retirement in England, he kept an active interest in promoting the region he knew so well. In his autobiography (Hose, 1927) he narrates how he wrote to the Rajah and asked for permission to show a map of oil seeps and some oil samples to the Shell Asiatic Petroleum Company. That was in 1907. Such was the interest raised by Charles Hose's data that a concession and lease was negotiated and granted to Shell by the Rajah in 1909. The same year, Charles Hose travelled back to Sarawak, to accompany Shell's geologist J. Erb and show him the locations of oil seeps in and around Miri. Within one year this led to the discovery of oil in Miri Well#1!

Charles Hose was one of the eminent pioneers in the establishment of the petroleum industry in Sarawak. For his services to Shell, he was apparently offered a significant sum of money, which he declined, preferring to receive a royalty on production –indeed a wise decision on his part! The signature and a portrait of Charles Hose while living in retirement in England (redrawn for this publication) are shown in Figure 1.3.1b and c.

Figure 1.3.1c: Portrait of Charles Hose, in later years (artwork by Setsuko Yoshida from a historical photograph)

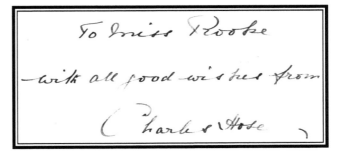

Figure 1.3.1b: Charles Hose's signature

1.3.2 Joseph Theodor Erb (1874-1934)

Joseph Theodor Erb (Figure 1.3.2a) joined the Bataafsche Petroleum Maatschappij (BPM) in 1900 as one of their first geologists. By the end of the first decade of his service, Erb was working closely with the Company's first Director General, Dr. H.W.A. Deterding. Erb achieved fame in 1910 when he successfully drilled the first oil discovery well (Miri#1) in Sarawak. The Miri field produced more than 80 million barrels of oil until it was abandoned in 1972.

Early in 1911 and barely back from Borneo, Erb was sent by Deterding on a mission to Illinois to assess the Bridgeport oil field. Meanwhile, the BPM's Central Geological Department was established in The Hague, and Erb was eventually called back to lead it. It was in his capacity as Chief Geologist that Erb wrote the *"Uniformity in Geological Reports"*, a company report establishing the first standard legend in the oil industry. Erb continued to take additional responsibilities, and in 1915, acting as chief negotiator for the Shell Group, he attempted to acquire the Healdton oil field in Oklahoma.

Erb had a keen interest in using new technologies for petroleum exploration. For instance, when the Eötvös torsion balance was proven to measure minor gravitational fluctuations, he immediately recognized the relevance of this instrument for locating buried anticlines. Soon after World War I, Erb commissioned torsion balances for the Shell Group and utilized them in the Egypt, Borneo, and Mexico. That new methodology led to the discovery of the giant Seria oil field, in neighboring Brunei.

In 1921, Erb was promoted to the rank of Managing Director of the Royal Dutch Shell Group, becoming a member of the Board from 1929 until his death in 1934. He was a member of AAPG until 1924.

Figures 1.3.2a: Portrait of Joseph Theodor Erb

1.3.3 Shell Staff Geologists (1910-1941)

Little information from Shell's internal company reports has survived with regard to the period before World War II. Destruction of these documents may have been part of the denial strategy of the Allies as the Japanese invasion loomed in 1941. From a surviving company report by P. von Schumacher, L.C. Artis and D. Gow on the Miri field, *The Geology and Prospects of the Miri Field* (February 1941; see figures 13.3a,b), we get glimpses of the enormous work carried out by the early geologists in Sarawak. We learn, for instance, that Joseph Theodor Erb issued the first geological map of Miri on July 16, 1910. We also learn the names of the first geologists associated with the appraisal and development of the Miri field: T.E.G. Bailey reported on well results and on the structure of the Miri anticline (1914-1915); R. Schider described the geology of the Miri oil field (1922); E. Braendlin reported on core drilling and gravity observations (1922); C.M. Pollok wrote on major tectonic features (1928) and issued a geological map of the Miri field (1931).

Between 1930 and 1934, the great economic depression drastically reduced the oil activities in Miri; in particular, exploratory drilling came to a temporary halt in 1931 and field geology was minimal, only to pick up slowly again from 1933 onward.

A.G. Hutchison discussed the Miocene-Pliocene unconformity (1933), D. Trumpy and S.T. Waite studied the Tukau Formation (1933); W. Sumner reported on his field work (1935); G.F. Wallis issued a memo on punch core holes (1936); and J. Schoo studied sub-thrust prospects (1937). With the pioneering work of R.W. Pooley (1937), A.M. Oosterbaan (1939), L.C. Artis (1940) and G.R.J. Terpstra (1940), the new correlation technique based on biostratigraphic analyses became routine from 1934 onward.

This flurry of activities was abruptly terminated at the onset of the war, with staff being repatriated.

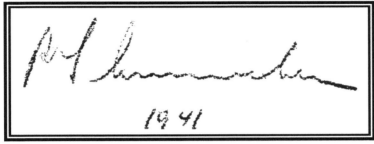

Figure 1.3.3a: P.Von Schumacher's Signature

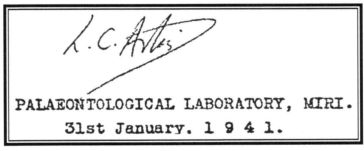

Figure 1.3.3b: L.C. Artis's Signature

1.3.4 Johann Karl Ludwig Martin (1851-1942)

Shell geologists working in Northwest Borneo frequently encountered fossils in their outcrop studies; these macrofossils were important to determine the ages of the sediments (in those early days, studies of macrofossils were more popular and micropaleontology was still in its infancy.) These fossil samples were usually sent to specialists in the Netherlands. BPM geologists sent their samples to Professor Dr. Karl Martin in Leiden. Martin, of German origin, was the first professor of geology at Leiden University, close to the BPM head office in The Hague. Martin became famous with his publication *Die Fossilien von Java* (1891), in which he described numerous new species of mollusks, corals and other fossils obtained from Miocene through Pleistocene sedimentary rocks of Java. He also proposed a novel method for dating sedimentary rocks. He counted the percentage of extinct species in fossil faunas and considered this to be a direct indicator of the age.

There are, as we know now, flaws with this method (due to relatively poor knowledge at that time of Recent mollusk faunas of southeast Asia, lack of differentiation between faunas from different depositional environments, variations in sample size, etc.), but Martin's work did help in the initial establishment of the relative ages of the Tertiary rocks in that part of the world.

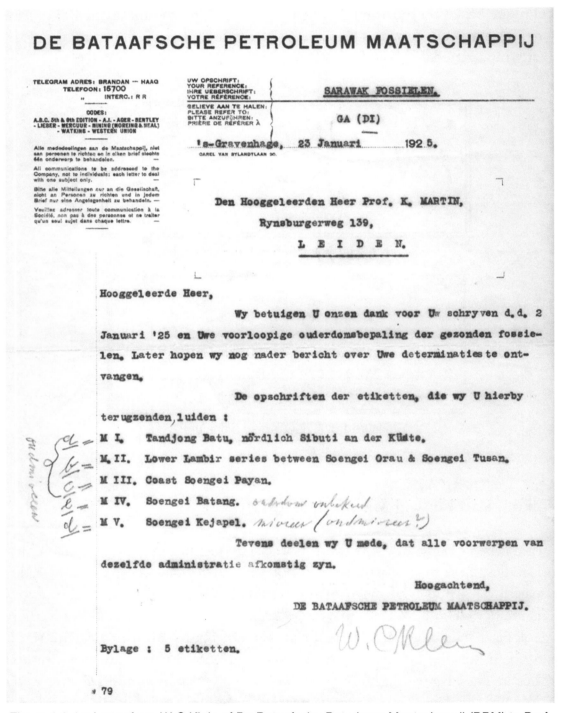

Figure 1.3.4a: Letter from W.C Klein of De Bataafsche Petroleum Maatschappij (BPM) to Prof. Dr. Martin regarding the samples from the Beraya-Tusan seacliffs.

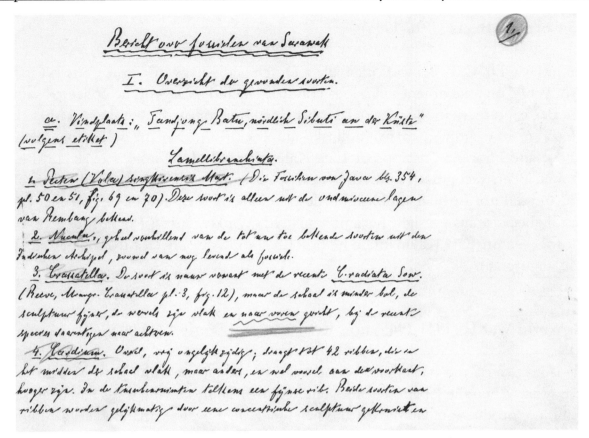

Figure 1.3.4b: First page of a hand-written and unpublished report by Prof. Martin (1925)

Martin received many samples from the Beraya-Tusan coastal cliff outcrops (south of Miri) and between Tutong and Muara (Brunei), as well as some core materials taken from wells in the Miri and Seria oil fields. The core samples usually yielded small specimens for analysis; moreover, they lacked a proper description of the stratigraphic section from which they were collected. Nonetheless, some of them formed the basis for newly described species, and thus proved to be valuable to science. Based on more recent investigations, it is possible to define the stratigraphic position for many of these early samples.

Martin knew little or nothing about the conditions under which the samples were collected by the BPM explorers: Only a few macrofossils were collected from the study outcrop, the samples were in small amounts, and even quite possibly specimens from multiple layers were mixed. Martin also had access to information about fossil collections at the Leiden museum and paleontological books on shells from the Indo-Pacific region. In 1891-92 he visited the Moluccas, where he also collected recent and fossil materials. He wrote many reports (often brief and hand written) on the BPM samples he studied, and sent copies of his reports to the company. Although Beets (1947: 40) comments that Martin's report about the Sarawak outcrops and Brunei went missing, all of his original reports (Martin, 1925, 1926, 1931, 1932) and the samples are preserved at the Netherlands Centre for Biodiversity Naturalis in Leiden. Figures 1.3.4a,b. show one of the letters sent by BPM to Martin and the first page of Martin's 1925 report.

1.3.5 Cornelis Beets (1916-1995)

Born in Java, Dr. C. Beets was originally a Dutch geologist working for Shell (formerly BPM). While on assignment in various countries including Egypt, he collected samples of modern and fossil mollusks. He later became Director of the Geological Museum in Leiden. He re-examined many mollusk samples of the Martin collection from the Miocene and Pliocene outcrops of East Kalimantan (Indonesia), Sarawak and Brunei. Although Beets recognized numerous new species, he made hand-written notes about them, and did not formally publish his descriptions of the Sarawak and Brunei faunas, probably because his eyesight became impaired (Winkler Prins, 1996). Much of Beets' work focused on East Kalimantan; he described only one species from Sarawak (Plate 4.5.4e, Figure 16).

1.3.6 World War II (1941-1945) and Rehabilitation Period (1946-1948)

Little is known of the oil activities conducted by the Japanese during their occupation of Sarawak. When the Japanese forces entered Miri, all producing wells had been sealed off, and the production equipment and machinery had been dismantled and shipped to Singapore. Nevertheless, the Japanese were able to trace back the missing equipment and to bring it back quickly to Miri. As access to oil was vital to the war effort, the Japanese with the support of skilled workers from Singapore managed to bring the field back to production; during the occupation period the Miri field produced some 700,000 barrels.

The Australian forces that liberated Miri in June 1945 encountered the oil wells on fire; immediate attention was given to the blazing wells, and the last of the fires was extinguished after three months.

Two years of effort were required to bring the field back on stream. Little geological work was carried out during this time. In 1948, M. Reinhard and E. Wenk from Basle University were appointed by Shell to review and document the geology of North Borneo. It took three years for their landmark report (*Geology of the colony of North Borneo*, 1951) to be prepared and published as the first issue of the Bulletin of the newly established Geological Survey Department, British Territories in Borneo.

1.3.7 The Geological Survey and Collaborative Geological Research (1949-Present)

Since its inception in March 1949, the Geological Survey Department, British Territories in Borneo was at the forefront of efforts to systematically map the geology of Sarawak, Brunei and Sabah. The Survey geologists produced detailed reports in the form of the Survey's *Memoirs* and *Bulletins* that still today form the basis of much of our present knowledge of the region.

Under the directorship of F.W. Roe, F.H. Fitch and R.A.M. Wilson, the Survey geologists pursued their titanic efforts to visit, map and analyze the geology of all remote areas of this huge territory, which remains poorly accessible even today. We are much indebted to field geologists, such as N. S. Haile, G. E. Wilford, P. Collenette, H.G.C. Kirk, E.A. Stephen, E.B. Wolfenden and others for the high quality of their reports.

Important geological reports about the Miri area include Wilford's *Geology and Mineral Resources of Brunei and Adjacent Parts of Sarawak, with Descriptions of Seria and Miri Oilfields* (1961), and Haile's *Geology and Mineral Resources of the Suai-Baram Area* (1962), as well as Shell geologist P. Liechti's and co-workers' *Geology of Sarawak, Brunei and the Western Part of North Borneo* (1960). In 1963, the Geological Survey Department, British Territories in Borneo became an independent branch of the Geological Survey of Malaysia, called the Geological Survey, Borneo Region, Malaysia.

Application of modern sedimentology to studies of Sarawak began in 1965 by the Shell Company, with the commission of a special study on the recent sedimentation of the coastal and offshore areas of Northwest Borneo. The study aimed to obtain a set of sedimentological parameters to allow the determination of depositional environments and the prediction of the patterns of sand distribution in the subsurface. The study also included an analysis of the foraminiferal and palynomorph content of the sediments. For this purpose, some 1590 sediment samples and over 1000 water samples were analyzed by Lalanne de Haut et al. (1968). The results of the study were published in an internal Shell report, *Recent Sedimentation in the Baram Delta and Adjacent Areas*, that described in detail 12 separate depositional environments for Sarawak and established a close correlation between depositional environments and distribution of major sand units. This subdivision of the depositional environments was applied to the Tertiary sections in exploration wells in NW Borneo, and further calibrated with log shapes and dipmeter data.

In 1969, the Geological Survey under Malaysian management was split into the Sarawak and Sabah divisions. The Sarawak division with headquarters in Kuching has been under the successive leadership of C. Kho Chin Heng, Chen Shick Pei, and A. Unya Anak Ambun, and is continuing its task of surveying the State of Sarawak. Since its foundation in 1974, PETRONAS and its subsidiary CARIGALI have complemented the tasks of the Survey by mapping and exploring all areas offshore Sarawak. PETRONAS geologists have played a prominent role in publishing their works, including the impressive book *The Petroleum Geology and Resources of Malaysia* (1999). A more recent publication, *Geology of North-West Borneo: Sarawak, Brunei and Sabah* (2005), by Charles S. Hutchison is also noteworthy.

1.4 The Miri Oil Field Story

1.4.1 Pre-discovery (up to 1910)

In his book *Fifty years of Romance and Research in Borneo* (1927), Hose provides the earliest writings about the oil potential in the Miri area. Quoting the diaries of Claude Champion de Crespigny, former Resident of the Third Division (Baram District), Hose relates: "31 July 1882. The celebrated earth oil at Miri found in about eighteen wells which some people dug some years ago in the hope of being purchased by possible buyers" and further "12 May 1884. I think the oil district near the mouth of Miri should be thoroughly searched and reported on". Given the curiosity of the Borneo tribes for all nature products, it is no wonder that the seeps of *minyak batu* (literally "oil from rock") would have been an object of special attention. Still today, local apothecaries in Miri will sell *minyak batu* as a remedy for stiffness and joint pain. It is thus entirely possible that since time immemorial, local people extracted oil at Miri for their private use.

While in Sarawak, Hose mapped the Miri area for oil seeps and even offered rewards to local people if they could show him new seeps. He then consulted with a visiting geologist who thought the oil seeps to be non-commercial. Later on, with the Rajah's approval, Hose arranged to dig a few crude test holes, each several feet deep, but encountered no success. He was then instructed to "stop wasting money" on drilling. Hose obeyed this order but continued to record oil seeps on his map, identifying 28 to 30 occurrences. It is unfortunate that Hose's map is not extant today.

1.4.2 Discovery and Early Production (1909-1919)

Much happened in the years 1909-1910! We have already seen how Hose was instrumental in attracting Shell to Sarawak, which led to a lease agreement signed in London in early 1909. Shortly thereafter, Hose together with Shell geologist Joseph Erb took trains bound for Moscow, where they embarked the Trans-Siberian Express to Vladivostok, and from there, they traveled by boat to Shanghai, Singapore, Kuching and finally Miri! G. Howell (1926) notes: "Between August and December in the year 1909, Dr. Erb carried out a rapid geological survey in the northern part of Sarawak…and was able to report the existence at Miri of a dome-shape, unsymmetrical anticline with a steep eastern flank and numerous oil shows". Erb left Sarawak in December 1909, only to return two months later, and he then started planning a location for the first well on Miri Hill. He refined his geological mapping in and around Miri; his first geological map of the area was included in an internal Shell report dated 16 July 1910 although, regrettably, no copies of the report or the map have survived. Drilling equipment must have been organized early in 1910 itself, for Miri#1 was spudded on 10 August 1910 (Figures 1.4.2a and b). It hit oil on 22 December 1910 at a depth of 452ft. The well initially produced some 83 barrels of light crude oil per day from a sand interval at the bottom of the hole that would later be called "No.1 sand". The well location proved to have been optimal, for this first well is prized as the longest producing well (62 years of production) and achieved the highest cumulative production volumes of any well in the entire Miri field, producing about 660,000 barrels of oil!

Miri#1 and all subsequent wells until 1924 were drilled using the old cable tool method, whereby a heavy bit with a blunt chisel, suspended in the derrick from a steel cable, was dropped repeatedly onto the bottom of the hole to crush the rock. After the bit had penetrated the formation for a few feet, it was pulled out, and the rock fragments in the hole (the cuttings) were removed with an open tube fitted with a valve at the bottom (the bailer). The oil found at the bottom of Miri#1 did not flow to the surface; it had to be pumped up, using a beam fitted with a revolving bull wheel. All these operations were driven by steam engines.

Following the success of Miri#1, Erb started planning for a second well on Miri Hill. He selected a location some 270 m northwest of the first well; the second well was drilled in April 1911, but, against all expectations, it did not encounter any oil. This was a clear indication that the apparently simple anticline seen on surface was more complex in the subsurface.

Figure 1.4.2a: Drilling well #1 in 1910 (courtesy of Sarawak Shell Berhad)

At the end of 1911, oil production from Miri#1 had reached 1950 barrels. Plans were laid out to erect a refinery, and for this purpose, a site called "Brighton" was selected near Tanjong Lobang. At the end of 1912, the refinery had handled 42,260 barrels, and one year later some 8 wells were producing together 195,500 barrels per year. As there was no port facility in Miri, oil was exported in sealed drums, carried offshore to anchored tankers by a fleet of small vessels. It was clear that such a method could not cope with the expanding oil production from the field, and a submarine pipeline was planned, to provide a direct route from the refinery to the tankers. In August 1914, a sea-line was laid down offshore the refinery at Tanjong Lobang. It consisted of 3 sections of 6" pipe, each 4,500 ft long. The total oil production during 1914 reached 483,825 barrels. As World War I started, in that year production from the Miri oil field continued at a remarkable growth, as there was crucial demand for oil to fuel the fleet of Great Britain's Navy and her allies. During the war years, the Miri field supplied some 2,152,500 barrels of oil. In order to meet the expanding demand, a better location for a larger refinery was sought. Following a detailed survey of the offshore area, a location was selected and in 1917 a new refinery and a new sea-line were commissioned at Lutong. At the end of 1919, some 75 wells had been drilled and the Miri field had already produced over 3.5 million barrels of oil.

Figure 1.4.2b: Coolies hauling a boiler up a hill in Miri, 1911 (courtesy of Sarawak Shell Berhad)

1.4.3 Ramp up to Peak Production (1920-1929)

World War I ended in 1918. The Miri Oil field continued its growth. During 1920-21, some 80 new wells were drilled, bringing the annual production to more than 1.5 million barrels of oil. At the same time two new sealines were laid out of the refinery at Lutong. The third sealine was 14,556 ft long, a world record at that time! In 1922, the yearly production doubled to 3.1 million barrels, and in 1924, it approached 4.4 million barrels, from 275 wells which had been drilled by then (Figure 1.4.3a).

Drilling by the cable-tool method was a lengthy process, sometimes taking more than a year for a well and the total depth was no more than 2000 ft. In 1923, Miri#193 was drilled as a deep test of the Miri Formation; it reached its total depth of 4,030ft five years later, fulfilling its objectives and penetrating the underlying Setap Shale. By 1925, the new rotary-drilling method was introduced, which significantly cut down the time required for drilling, and also allowed for cores to be retrieved. This method relies on a tooth-bit in a continuous circular motion to grind the rocks at the bottom of the hole. Drilling fluids (mud) are injected through the drill pipe down to the level of the bit, and as they circulate between the pipe and the borehole, they bring up the cuttings to the surface. This new technique was used to drill Miri#330, starting in August 1926; it reached a total depth of 4,605ft in July 1927. With time, drillers mastered the rotary drilling more efficiently: Miri#358 was spudded in January 1927, and reached a total depth of 2000 feet in April of the same year. Although most wells were drilled by the rotary method, the cable-tool method was continued as a parallel technique until 1930; one of the last wells to be drilled by this method was Miri#538, which began in January 1930, and reached a total depth of 1500ft in April of the same year. In 1927, some 88 new wells were drilled, which proved a record year in the development of the Miri field. By then, yearly production levels had increased to over 5 million barrels of oil (Figure 1.4.3b).

The years 1928 and 1929 marked peak production, with respectively 5,475,000 and 5,548,000 barrels of oil being produced. During these peak production years, some 15,211 barrels per day were produced from the field. In these two years 110 new wells were drilled, including the ultra-deep Miri#498, which penetrated 6,180ft of sediments, including the entire Miri Formation and some 1760ft of Setap Shale. At that time, the field had been densely covered by drilling (Figure 1.4.3c), with most wells drilled in circa 300 ft spacing. In 1929, with the Miri field at its peak production, a new oil field was discovered by Shell in neighboring Brunei: the Seria field would prove to be of much larger size, and would compensate for production decrease of the Miri field.

Figure 1.4.3a: Miri#1 and nearby derricks in a picture from Charles Hose (1927)

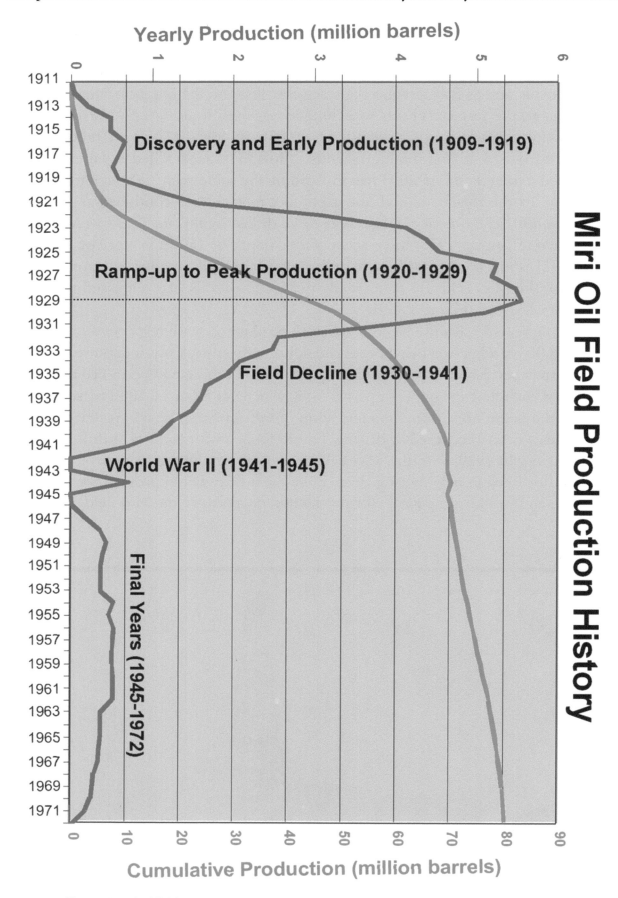

Figure 1.4.3b: Miri field production history (adapted from Sarawak Shell Berhad, 1990)

1.4.4 Field Decline (1930-1941)

The year 1930 still marked a healthy production, some 5.1 million barrels of oil for the Miri field, but it plunged dramatically the following years: in 1931 the field yielded some 3.9 million barrels, and in 1932, production was down to 2.5 million barrels. Admittedly, the number of wells drilled during these years of worldwide economic crisis was also much reduced: 32 wells in 1930, only 5 in 1931, and none at all in 1932, but the fact was that the producing wells (Figure 1.4.4a) were rapidly depleting the field.

During 1933-1934, yearly production was still above 2 million barrels, but it fell below this threshold in 1935. At that time, Shell was developing the Seria oil field; therefore, oil from Seria was piped to the Lutong refinery, keeping it at capacity and compensating for the dwindling production from the Miri field.

During 1936-1937, yearly field production from Miri was just above 1.6 million barrels of oil, it was barely 1.5 million barrels in 1938; and by 1939 it was down to 1.2 million barrels. Some 25 wells were drilled during these two years, looking for new oil pockets in the field, but encountered limited success.

This was a period when new geologic methods were tested: Micropaleontology was introduced in an effort to resolve the geological subsurface model of the field, which was extremely complex because of the numerous faults criss-crossing the field. It had always been much of a puzzle how to correlate sandstone units from one well to another: finding sandstone at the same depth in two neighboring wells did not necessarily indicate that they were the same sand units because faults might have reset the order of the sandstone layers. Initially, geologists thought that they could identify individual sand units based on the gravity of the oil it contained: shallower sandstone harbored the heaviest oil (probably due to biodegradation), and the deeper, older sandstones contained lighter oil.

Micropaleontology or the study of microfossils proved to be a much more scientific method of correlating rock units, and its systematic application, beginning in 1937, brought fundamental changes to the model of the subsurface geology of the Miri field. A number of shale units proved to be characterized by a unique assemblage of microfossils, which allowed them to be identified and correlated with great certainty from well to well.

Thus, the years leading to World War II were marked by a frenzy of activities in trying to create a better geologic model for the Miri field, in the hope to identify new areas of hydrocarbon prospectivity. Shell company reports during this period focused on understanding the hydrocarbon trapping mechanism of faults and also looking at deeper targets in the field that had not yet been penetrated. All this research was strengthened by micropaleontological studies. *The Geology and Prospects of the Miri Field*, a landmark Shell report was issued in February 1941, co-authored by a geologist (P. von Schumacher), a micropaleontologist (L.C. Artis), and a production engineer (D. Gow), which summarized the status of geologic knowledge of the Miri field at that time, and made recommendations to explore further areas in the northern part of the field, as well as drilling deeper prospects. This report was a vanguard of multidisciplinary work that is often claimed to be a new trend nowadays. As the prospect of war was looming, none of the recommendations by von Schumacher et al. were carried out. Before the Japanese invasion in December 1941, yearly production from the Miri field was barely 1 million barrels in 1939 and about 700,000 barrels in 1940.

Figure 1.4.3c: Miri#23, drilled next to Krokop cemetery, at a time when the area was still a tropical forest (courtesy of Sarawak Shell Berhad)

Figure 1.4.4a: Hillside wells, Miri Field (courtesy of Sarawak Shell Berhad)

1.4.5 World War II (1941-1945)

The landing of the Japanese forces on the Norwest Borneo coast was primarily to capture the oil fields of Miri and Seria. As part of the "denial policy" adopted by the Sarawak, Brunei Government and Shell, all producing wells were to be sealed and all means of oil production were to be dismantled. However, the invasion occurred earlier than anticipated, and whilst the Miri oil field had been shut down, the production equipment had barely left Sarawak for Singapore when the Japanese forces landed at Tanjung Lobang on 16 December 1941. Following the fall of Singapore, the Japanese forces were able to return to Miri the same drilling and production equipment that had been shipped out. Skilled technicians that had left and accompanied the shipment were also forced back to Miri and had now to work for the Nenryo Hai Kyusho, or the Oil Supplying Service, under the Japanese military (Sarawak Shell Berhad, 1990).

It took great effort and time for the Japanese to restore reasonable production levels from the Miri field. During the last 2 years of the war, they managed to produce an estimated 710,000 barrels of oil. War-time oil from both Miri and Seria fields, which is estimated to total 11,498,000 barrels of oil, was refined at Lutong. It was at that time that the Japanese forces constructed the Lutong airstrip.

In June 1945, as the Australian 9th Division took Seria and Miri back from the Japanese, there was utter devastation of the oil producing facilities and wells were on

Figure 1.4.5a: Fires in the oil fields, 1945 (courtesy of Sarawak Shell Berhad)

fire; in some cases pits filled with flaring oil had been dug around the wellheads to impede access (Figure 1.4.5a). It was the Japanese turn of implementing a denial policy.

1.4.6 The Final Years (1945-1972)

Major rehabilitation efforts were carried out shortly after the end of the war in 1945. One of the first tasks was to bring the burning oil wells under control, which was achieved in September 1945. The storage tanks at Lutong were rebuilt in December 1945, as was the Seria-Lutong pipeline. Oil started flowing again, but initially from Seria only.

Late in 1946, some low levels of production (about 20,000 barrels for that year) were again achieved from the Miri oil field. The refinery at Lutong had to be rebuilt and was ready to be commissioned at the same time as the Seria-Lutong trunkline #2 was completed. The same year, drilling operations restarted and met with some success, bringing a good omen to the population; the first drilled well reached a total depth of 3,346ft and produced a steady 420 barrels of oil per day.

From 1947 through 1949, yearly production levels kept a steady increase from 182,500 to 365,000, and to 438,000 barrels of oil respectively, but these figures were modest by pre-war time standards, and the Miri field never gained its former glorious production levels. Miri #1, which had remained undamaged during the war, was rehabilitated on March 12, 1947 and started producing initially 9 barrels of oil per day (Figures 1.4.6a and b).

In 1954 yearly production approached 530,000 barrels of oil, but the down-ward trend was inevitable and in 1956 the field was producing less that it did in 1916! The introduction of electrical pumps in 1955 had little effect on the overall production from the field. New hydrocarbon resources had to be located elsewhere in Sarawak.

Figure 1.4.6a: Miri Well#1 in the 1950s; note the 3-legged derrick in the background, probably Miri#42 (courtesy Sarawak Shell Berhad)

From 1953 through 1960, Shell explored aggressively the inboard areas of Sarawak in the hope of discovering a new oil field. However, drilling of many prospects, including those at Suai, Mukah, Balingian, Tatau, Lambir Hills, Sungai Madalam, Pasir, Bakam, Selungun and Semilajau brought no success to the company. In 1954, Shell obtained a mining lease covering the entire Sarawak shelf, which in the medium term proved to be a prolific ground to explore for oil. From 1962 to 1970, as the Miri oil field was nearing the end of its production life, a series of hydrocarbon discoveries were made in the offshore: Temana field (1962),

Baram field (1964), West Lutong and Tukau fields (1966), Baronia and Baram B fields (1967), Betty field (1968), Bakau and Bokor fields (1970).

From 1956 to 1962, the Miri field still produced a good half a million barrels of oil per year; the production was, however, reduced to 300,000 barrels per year by 1967, and continued its downfall thereafter. New techniques to boost production, such as water injection were tried, but with little success. Testing of potential extensions of the field towards Tudan and Luak Bay also proved negative. As the field was in its final decline, the population of Miri was increasing and larger parts of the field polygon were now occupied as residential or business areas. Drilling plans in these areas were curtailed as they created hazards to the population.

On the 20th October 1972, the Miri field was shut down as oil production levels had dwindled and the average watercut per well had reached 90%. At that time, only 70 wells out of 623 were pumping an average 275 barrels of oil per day; only 4 wells produced more than 10 barrels of oil per day (Figure 1.4.6c). The last producing well was Miri#549, which yielded 19 barrels of oil and 300 barrels of water per day. Dr. Erb's well Miri#1, now affectionately referred to as the "Grand Old Lady", was still producing about 7 barrels of oil per day when it was finally shut down (Figure 1.4.6d).

A small sub-block of the field was retained for enhanced oil recovery (EOR) experiments, and two wells (Miri#611 and Miri#612) were drilled for that purpose in 1979. However, as the EOR results were not encouraging, this sub-block was

Figure 1.4.6b: Walkway to Miri Well #1 (courtesy of Sarawak Shell Berhad)

Figure 1.4.6c: A typical Miri oil field scene in 1970 (courtesy of Sarawak Shell Berhad)

relinquished in 1981. At the time, PETRONAS erected a number of concrete markers at the site of key wells on Canada Hill (Figure 1.4.6e).

Interestingly no seismic was ever acquired over the Miri field and wireline logs (mostly induction logs or gamma-ray logs) were obtained only for the 34 post-war wells.

Figure 1.4.6d: Grand Old Lady in retirement (courtesy Sarawak Shell Berhad)

Figure 1.4.6e: PETRONAS' marker for Miri#42, at a short
distance from Miri#1

1.4.7 A Concluding Note from Dr. Erb

In 1916, writing from the head office of Shell, Dr. Erb took a dim view of the prospectivity of other structures onshore Sarawak. His letter states: *"The writer, who did the pioneering work in the Baram District, watches with the greatest interest the development of the Miri field ... He always laid the stress on the fact that the Miri anticline is only of middle size and that a moderate oilfield would more likely be developed than a very rich pool ... It must, however, be borne in mind that the other places in Sarawak where oil-seepages occur, offer less prospects than Miri itself. The anticline of Bulak-Malang-Setap has been carefully examined and tested by borings without result. The Buri anticline might become a small producer, but may also be barren, being constituted of rocks low down in the scale of formations ... A most far-reaching conclusion follows from this statement, viz- Miri might remain the only producing field of Sarawak, but even if another pool is developed it will doubtless be less rich, more difficult of access and the exploitation will be still more expensive than in Miri."*

This was, indeed, a visionary and prophetic statement by J. Erb, as no other field has been developed onshore Sarawak till today. (Note that the technology to explore and drill offshore was non-existent at the time of Dr. Erb; therefore, he did not comment on the offshore possibilities.)

Chapter 2

The Miri Oil Field in its Geological Context

2.1 The Record of Rocks

Standing in front of any cliff face around Miri, one will notice that the rocks are layered and that individual layers or beds vary in thickness, color, grain size, etc. Bedding surfaces usually correspond to minor breaks in sedimentation. On closer inspection, one will also see that some beds are homogeneous, massive sandstones or shales; some beds have preserved certain indications of the way the sediments were deposited (such as current scours, ripple marks), and some beds exhibit internal heterogeneities due to burrowing activities of animals at the time of deposition.

One of the geologist's primary tasks is to record in detail, at scale on a log, the make-up of outcrop rocks. Each outcrop offers only a partial view of the full rock sequence; therefore, the geologist has to piece together observations from various outcrops to build the big picture of the succession and composition of rocks in a given area. For the purpose of scientific analysis and communication, geologists usually divide sedimentary rocks into Groups, Formations, and Members, and assign specific names to them. For example, the thick sandstone sequence seen on Canada Hill is called the Miri Formation. Geological maps show the outline of the various rock units, each depicted by a different color or pattern (Figure 2.1a and 2.2a).

A significant step in the interpretation of sediments is reached when each bed or a succession of strata can be ascribed to a particular depositional environment. Sedimentological and biological studies of modern environments provide a reference framework for such an interpretation. In the Miri area (both onland and offshore), depositional environments have remained roughly similar over the past 20 million years, although they have shifted locations as land areas previously under water have been uplifted. Hence, the study of the present-day continental, coastal and marine environments acts as a guide for the interpretation of sedimentary structures and fossils found in outcrops. The main coastal depositional environments are related to deltas (such as the Baram Delta), and to tidal-influenced estuaries and embayments (such as Brunei Bay). Fully marine environments are found on the shelf, at water depths up to 200m, away from deltas. Near-shore (proximal) environments (0-50m water-depth) have a characteristic distribution of biota, with large and thick-shelled bivalves and gastropods, and other light-dependant creatures, such as corals; because strong currents affect shallow seas, the shells are often tumbled, abraded by sands and broken. In deep offshore (distal) environments (50-200m waterdepth), much lower energy conditions allow the shells to be better preserved as they are often embedded in mud.

While logging an outcrop, the geologist tries to decipher how depositional environments varied through time and space as various sedimentary units were deposited. The geologist looks for patterns in the depositional history in order to determine the relative changes in sea-level and the stratigraphic surfaces that separate different depositional environments such as marine flooding surfaces (an abrupt deepening in marine sedimentation) or sequence boundaries (an abrupt shallowing of depositional environments, often associated with a significant time gap).

In the course of geological time, rocks suffer tectonic deformations, as can be readily seen in outcrops along the Miri Airport Road (see chapter 3.4). Faults offset rocks in different orientations, which can be mapped and related to tectonic episodes in the region, such as extension in which rock blocks collapse along normal faults, or tectonic compression by which rocks are uplifted along reverse faults, thrusts, and folds (see chapter 3.5).

Driving along the Hospital Road in Miri, one will notice that sedimentary layers have been tilted at high angles, and subsequently eroded to form a terrace; horizontally bedded white sands overlay the tilted rocks (see chapter 3.6). This sort of contact between rocks is called an "angular unconformity," which indicates three subsequent cycles in the geological history of the area: (1) the first cycle corresponds to the deposition of sediments as flat beds on sea floor; (2) the second cycle corresponds to the tilting and erosion of the sediments, and (3) the third cycle corresponds to renewed deposition of sediments above the unconformity.

In a succession of horizontally bedded rocks, one can reasonably assume that the sediments at the base are older than those at the top. This is a relative dating of sediments, but other equally important questions need to be answered: What is the age of the sediments (how many thousands or millions of years ago did they form)? Were they deposited on land or under the sea? The geologist answers these questions by analyzing samples of sediments and by looking at diagnostic features preserved in the rock. For example, coarse pebble successions indicate deposition by rivers, and specific fossil shells indicate a marine environment. To unravel the actual ages of sediments, geologists use various techniques; one of the most common is to look for age-diagnostic microfossils. As organisms have evolved in the geological past, their shapes have changed. Each geological epoch is characterized by a unique set of fossils, which can be used to date the rocks in which the fossils are found (see chapter 4.1.3).

Radioactive decay of particular elements present in minerals also provides a tool to date geological events and rocks. Radioactive elements with varying number of neutrons (isotopes) have a constant rate of decay (loosing excess neutrons) through time; this rate of decay can be calculated for various elements. By knowing the radioactive decay rate and the amount of parent and daughter elements in a rock sample, scientists can determine the ages of rocks. Carbon-14 dating provides age estimates up to 50,000 years and is extensively employed in archeology. Potassium-argon, rubidium-strontium, and uranium-lead dating methods provide ages in millions and billions of years in the rock record. These methods are applied to dating of crystalline (volcanic, plutonic and metamorphic rocks), not sediments. In sedimentary successions, scientists sample volcanic tephra for radiometric dating. Decades of research in various parts of the planet have enabled geologists to construct the Geological Time Scale (Figure 2.1b), which is constantly being updated with new data by the International Commission on Stratigraphy (http://www.stratigraphy.org)

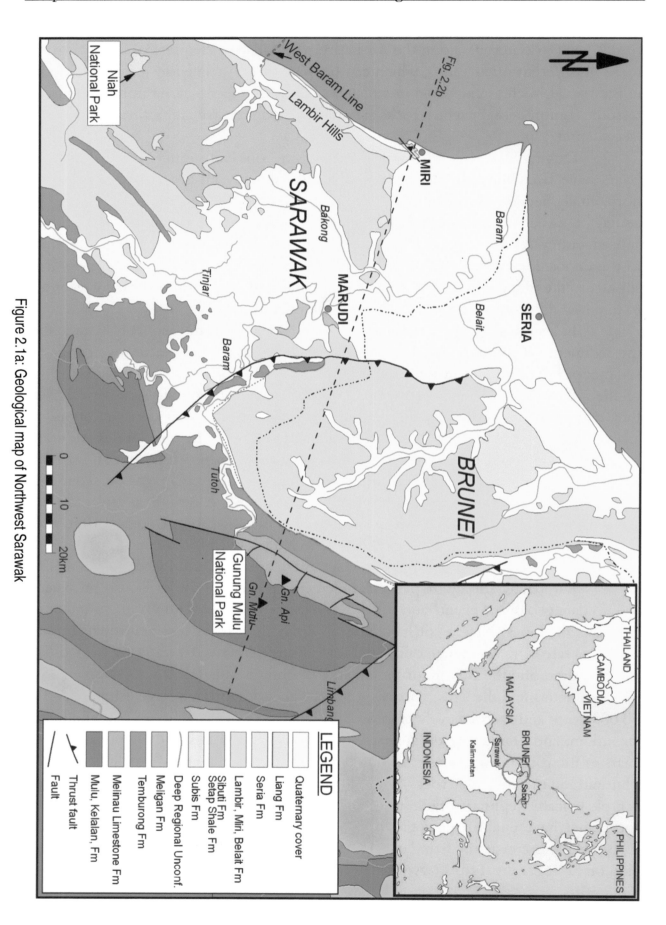

Figure 2.1a: Geological map of Northwest Sarawak

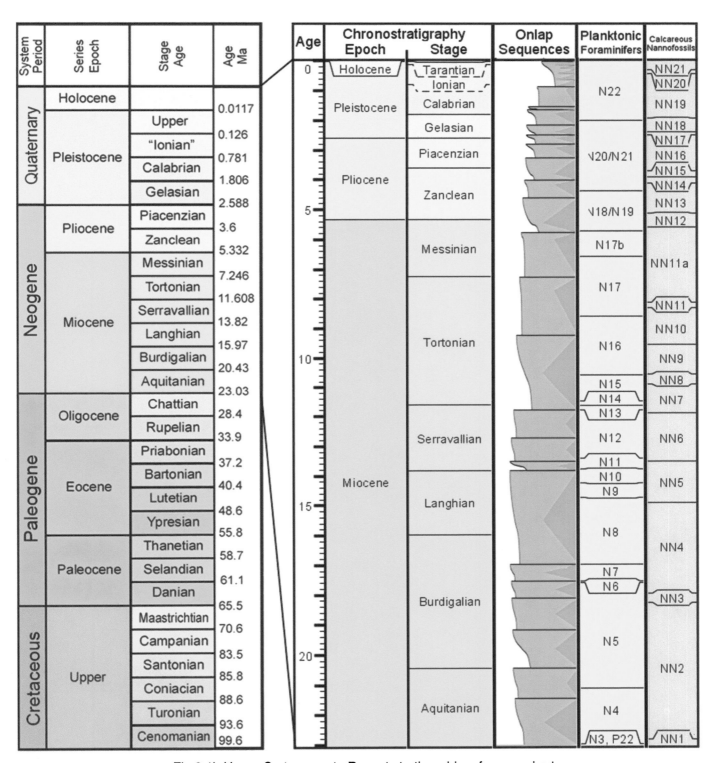

Fig 2.1b Upper Cretaceous to Recent stratigraphic reference chart

2.2 Tectonic Setting of Sarawak

Sarawak has an area of 124,450 square kilometers (48,050 square miles) and its major towns are Kuching (capital city), Sibu, Bintulu, and Miri (Figures 2.2a and 2.2b). Based on surface geological mapping, Haile (1974) divided Sarawak into three major zones separated by tectonic lines (suture zones), and each is named after a major town (Figure 2.2b):

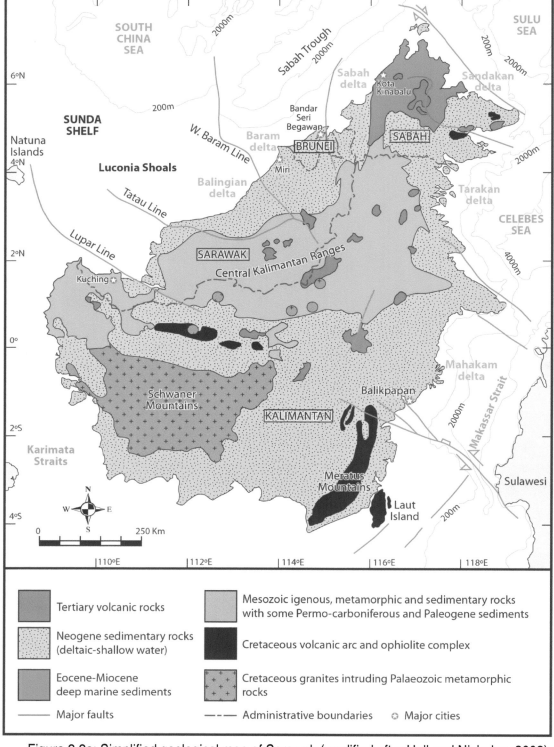

Figure 2.2a: Simplified geological map of Sarawak (modified after Hall and Nicholas, 2002)

(1) The Kuching Zone lies to the southwest of the Lupar Line (suture zone) and includes Paleozoic meta-sedimentary rocks (schists and phyllites) overlain by volcanics and sedimentary rocks of Mesozoic age and Tertiary clastics;

(2) The Sibu Zone of central Sarawak lies between the Lupar Line and the Tatau-Mersing Line (suture zone) and consists of Late Cretaceous to Eocene deep-water Rajang Group overlain by Oligocene-Miocene clastic sediments with coal deposits at the base (the Nyalau Formation); and

(3) The Miri Zone of northern Sarawak includes post-Eocene, mainly shallow marine and deltaic sediments (Tan et al., 1999).

Overall sedimentation becomes younger from the Kuching Zone in the south to the Miri Zone in the northwest, indicating the migration of basins through time as various tectonic events affected the region (e.g., Hutchinson, 1989; Hall, 1996).

To better understand these tectonic zones and suture zones we need consider the tectonic setting and evolution of the region (Figure 2.2c). Here we adopt the tectonic interpretation given by Madon (1999) adding some interpretations of our own.

The southern part of Borneo or Kalimantan continental block is part of Asian Sundaland that consists of Paleozoic rocks. The Lupar line is considered to be a paleo-suture zone between the Kuching Zone (part of the Kalimantan continental fragment) and the Rajang Sea (a seaway that separated Kalimantan from China during the Mesozoic). During the Paleocene, a subduction zone developed along the Lupar Line extending to central Kalimantan; the products of this subduction-accretion complex constitute the Rajang Group of rocks, present as a mountain range in the Sibu Zone. It is believed that the Rajang subduction was associated with the drift of a continental block (called the Luconia block) from China and its accretion to Borneo. The initial collision ("soft collision") probably took place in the Late Eocene and in an oblique manner from northwest to northeast. From the Late Eocene through the Middle Miocene, the paleogeography of Sarawak was controlled by tectonic convergence and the resultant uplift of the Rajang Mountains and sea-level changes resulting in fluvial to marine

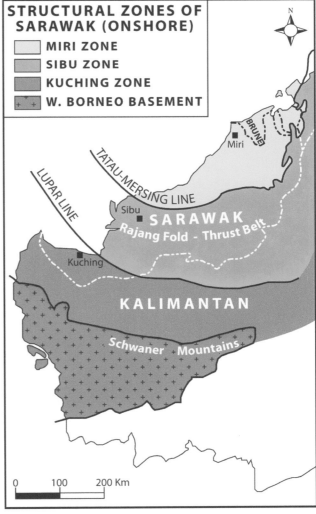

STRUCTURAL ZONES OF
SARAWAK (ONSHORE)
MIRI ZONE
SIBU ZONE
KUCHING ZONE
W. BORNEO BASEMENT

Figure 2.2b: Tectonic zones of Sarawak (modified after Leichti et al., 1960; Madon, 1999)

sedimentation. In the Middle Miocene, the "hard collision" of Luconia with Kalimantan continental block put an end to deep marine sedimentation and further elevated the Rajang Mountains in central Borneo. Both the Late Eocene "soft collision" and the Middle Miocene "hard collision" are reflected by major regional unconformities (hiatus in sedimentation as a result of uplift and erosion) in the sedimentary record of Sarawak. After the Middle Miocene, a huge deltaic system – the Champion and (paleo) Baram delta complex– was superimposed on the earlier marine sediments of Sarawak. Over time, this deltaic system has prograded seaward and has supplied a wedge of clastic sediments about 15 km thick. The whole sedimentary sequence has undergone syn-sedimentary growth-faulting during Late Miocene and Pliocene times. During the Pliocene, rock formations in Sarawak experienced a compressional deformation resulting in a series of folds and reverse faults now visible onshore. The origin of this compressional phase has not been definitely resolved; it was probably a combined effect of (1) subduction along the North Borneo Trough and (2) transpression generated from strike slip motions in the Indo-China Block across the South China Sea to Sarawak. The compression probably caused an inversion tectonics, reactivating the previous normal faults as reverse faults and folds (Morley et al., 2003). The Miri anticlinal structure bounded by reverse faults and adjacent to normal faults (possibly growth faults) epitomizes the structural evolution of Sarawak over the past 15 million years or so.

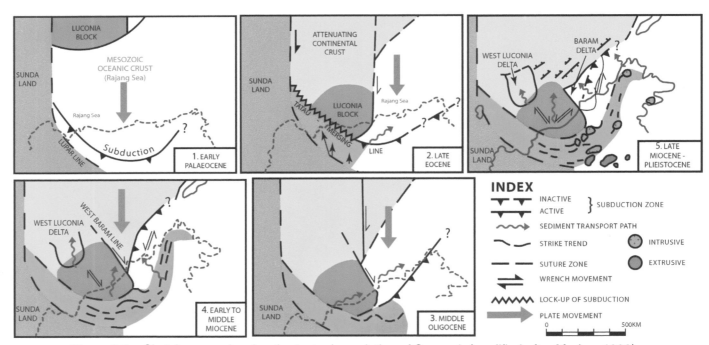

Figure 2.2c: Sketch maps showing the tectonic evolution of Sarawak (modified after Madon, 1999)

2.3 Geologic History of Northern Sarawak

Geological history is measured in millions of years and divided into periods and stages. The geological timescale and rock formations of northern Sarawak, relevant to Miri, are shown on Figure 2.3a. The stratigraphic record in our area covers the last 65 million years of Earth's History (Cenozoic Era). While most of the rocks are sediments deposited below sea-level, more recent sediments have been deposited onshore, on fluvial or alluvial plains. In other words, the geography of the region has changed over time. About 66 million years ago, when the dinosaurs were wiped out in an environmental cataclysm, the Rajang Mountains did not exist and Borneo was a much smaller island. A geological cross-section from Mulu to Miri (Figure 2.3b) shows the architectural arrangement of the various rock formations, and the overall younging of the sediments from south towards north.

The geological history of Sarawak can be subdivided into a number of phases described below.

2.3.1 Paleocene to Middle Eocene Times (65-40 Ma)

The whole region of northern Sarawak was once part of a deep marine basin, becoming shallower southwards, way beyond what is today Mount Mulu; none of the mountains that now make up northern Sarawak existed at that time. Muds, sands and pebbles were deposited in this deep sea; they have been preserved as shales, sandstones, and conglomerates that crop out on Mount Mulu and are called the Mulu Formation, part of the Rajang Group.

Mountains form when different geological terrains or tectonic plates move towards each other and collide. Such continental collision closes the oceanic basin that existed between the converging terrains or plates. Therefore, some of the deep marine sediments are uplifted, deformed and preserved in the form of mountains.

The Rajang Group was part of a larger sedimentary (accretionary) prism that extended all along Sarawak; while very deep marine conditions existed in today's Miri area, the seabed became shallower towards the south and formed the submarine slope of the Rajang Group. In fact, these slope-prism sediments formed a submarine mountain, as the oceanic crust under the Miri area was being subducted below the continental crust of central Borneo. Being muddy and water-rich, the sediments of the Rajang Group were easily deformed, piling upon each other and thus forming the slope that delineated a giant sedimentary prism. It is estimated that the deep marine shales deposited in front of the Rajang Group extend underneath Miri at a depth of 7-8km, and continue further into the offshore.

A major mountain-forming process occurred at about 40 Ma, presumably as a result of a "soft" collision between Luconia and Borneo. The Rajang sedimentary prism was then uplifted, folded, and underwent erosion. The corresponding unconformity in the rock record is called the Sarawak Unconformity.

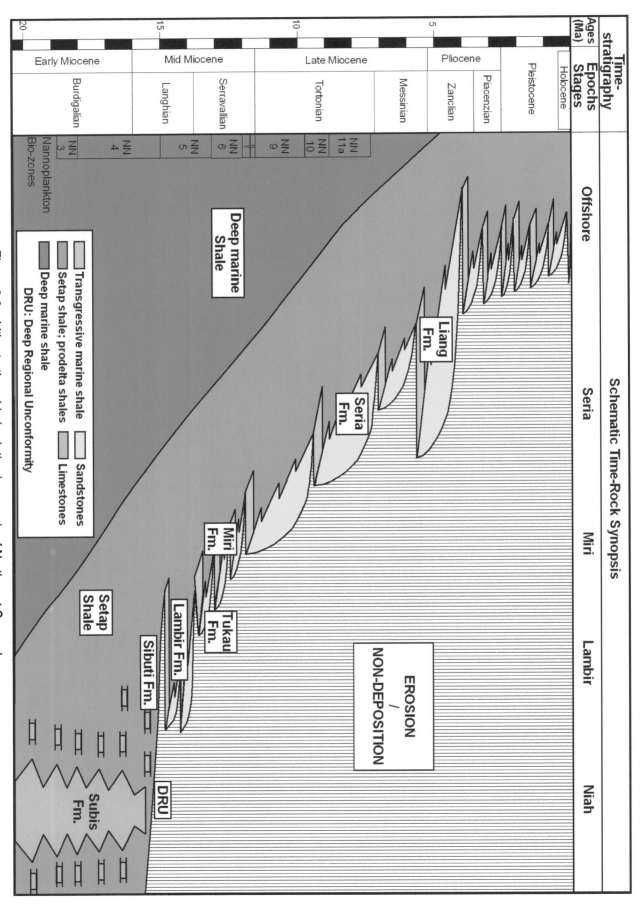

Figure 2.3a: Lithostratigraphic (rock-time) synopsis of Northwest Sarawak

2.3.2 Late Eocene to Middle Miocene Times (40-15 Ma)

On the higher submarine parts of the Rajang sedimentary prism, marine algae, corals and a host of other marine creatures were living in warm shallow waters; these organisms were all characterized by skeletons of calcium carbonate. Over time, the accumulation of these organic and lime-rich particles formed limestones, which can now be observed in various parts of Sarawak. The Melinau Limestone Formation, in which the Mulu caves were later developed, was deposited in this manner (see chapter 3.16).

During the early Miocene (20-15 Ma), the whole region, including the area of Miri, was again under deep marine waters; the Temburong Shale Formation was deposited during this time.

At about 17 Ma, the deep marine basin that had until then covered much of northern Sarawak became shallower towards southwest, and coral reefs started building up. Most of these reefs are today buried deep below the South China Sea, where they form the Luconia Platform. One of the southernmost reefs is exposed, forming the mountain of Batu Niah (here these limestones are called Subis Formation), an area famous for its cave system and valuable archaeological remains (see chapter 3.15).

An important geological event took place during the Middle Miocene that would have a major impact on the subsequent stratigraphical-structural evolution of the whole region: the Luconia Block, an Asian continental fragment that had drifted southward, completed its "hard" collision against the crust of central Borneo. This event is expressed in the sedimentary record by the Deep Regional Unconformity and signals the uplift and initial erosion of parts of central and northern Borneo. The West Baram Line, which today extends along the southwestern end of the Lambir Hills and continues offshore in a northwest-direction, probably became active at that time as a major strike-slip fault zone.

2.3.3 Middle Miocene to Pleistocene Times (15 Ma-12,000y ago)

Deep marine conditions persisted in northern Sarawak during the Middle Miocene: the Setap Shale Formation deposited at that time consists of a thick, monotonous succession of dark clays and shales with minor intercalations of siltstones and thin sandstone beds. Typically, Setap shales accumulated very rapidly and retained much of their interstitial water; this trapped water made the formation particularly ductile and created instabilities in the overburden sediments. Soft sediment deformations and mud volcanoes are some of the manifestations of the overpressures affecting the Setap Shales and its overburden. It is estimated that this formation extends down to some 5 km below Miri.

The upper part of the Setap Shale has higher carbonate content and is identified separately as the Sibuti Formation. Marls and thin limestones, often rich in fossils, characterize this formation. The contact between the Sibuti Formation and the overlying sandstones of the Lambir Formation is a regional unconformity (see chapter 3.12.5).

During the Middle to Late Miocene, the Rajang Mountains emerged from the sea and were then subjected to erosion by river systems. These streams cut through the exposed sandstones of the Rajang Formation and transported the sand grains westward, depositing them in the form of sand-rich deltas and embayments. The Lambir and Miri Formations are records of these shallow marine and estuarine environments dating back to 14-9 Ma. At that time, the area of Miri barely started to emerge from the sea.

Overlying the Setap Shales and Sibuti Formation, the Lambir, Tukau and Miri formations are sand-rich sediments that grade laterally into one another. The Lambir Formation, described from its type locality in the Lambir Hills, is an alternation of shallow marine sandstones and shales with a distinctive calcareous admixture, which may be locally absent (Liechti et al., 1960).

Paleontological work carried out in the late 1930s by Shell geologists led to the recognition that while shallow marine and estuarine sandstones were deposited in the Lambir area, the Miri area was the site of deep marine shale sedimentation. Based on deep wells in the Miri field (Miri#204, 290 and 330), fossil microfauna from these thick shales (1000-2500 ft) were recognized as the 'Loxostoma-1 Biozone", forming the lower part of the Miri Formation. The age of the Lambir and lower part of the Miri formations is unclear but likely ranges from the late Middle to the Late Miocene (14-10 Ma).

The Miri Formation (4500-5000 ft thick) proper was the productive interval of the Miri oilfield. It consists of seven separate, well consolidated, fairly hard, mostly fine grained and moderately porous sandstones, each about 500 ft thick, which are separated by shales or sandy shale beds each about 50-200ft in thickness. As the sand units of the Miri Formation were accumulating on the seabed, the pressure exerted by these

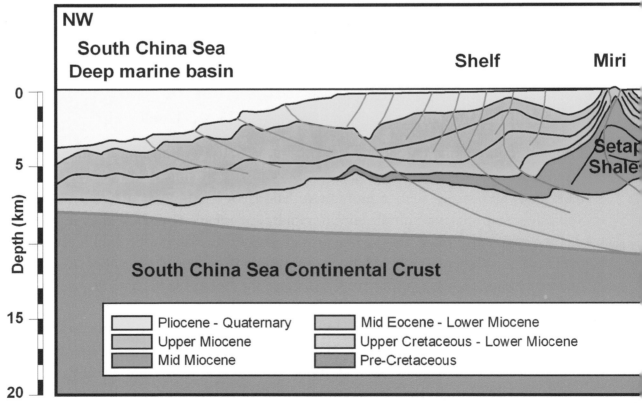

Figure 2.3b: Geological cross-section from Mulu to Miri (Modified after a Shell internal report by A. Gartrell & J. Torres); location of cross-section is indicated on figure 2.1a.

sediments on the underlying, water-rich Setap shales induced a subsurface retraction and flow of the shales. In turn, this resulted in syn-depositional deformations within the Miri sandstones, such as growth normal faults, block rotations and internal unconformities (see chapters 3.8 and 3.12.5).

Merely based on lithological criteria, it is impossible to separately identify each of the seven sandstone units. Shell geologists discovered that shales separating the sandstone units were characterized by specific foraminiferal fauna. The distinctly fully marine character of these shales, in contrast to the sandstones, led to the interpretation that each of the shale units corresponded to a phase of marine transgression (Figure 2.3.3a).

While fossil foraminifera allow the separate identification of the shale units, they do not constrain the age of the Miri Formation. Based on regional correlations, the age of the Miri Formation is widely thought to be Middle to Late Miocene (13-9 Ma).

The Tukau Formation is a lateral equivalent of the Miri Formation, developed between Miri and Lambir, and consists of a succession of deltaic-paralic, poorly consolidated sandstones and sands, alternating with soft clays, with frequent finely dispersed lignitic material or thin lignite layers (Liechti et al., 1960).

Offshore Brunei, further to the northeast, a large delta system developed as the Miri Formation was deposited; this Champion Delta reached a width of some 300 km and accumulated over 10 km of sediments during the Late Miocene.

Overlying the Miri Formation along an unconformable contact, the Seria Formation consists of a series of soft, thin laminated layers of clays, sandy clays and

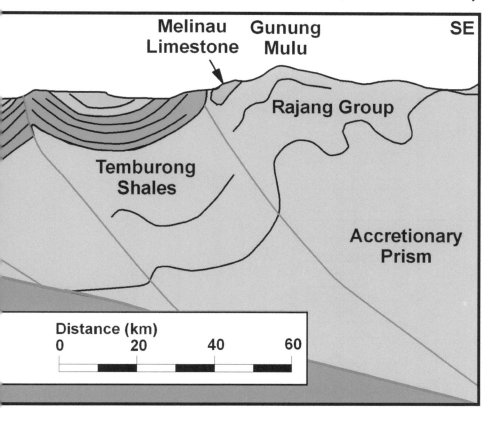

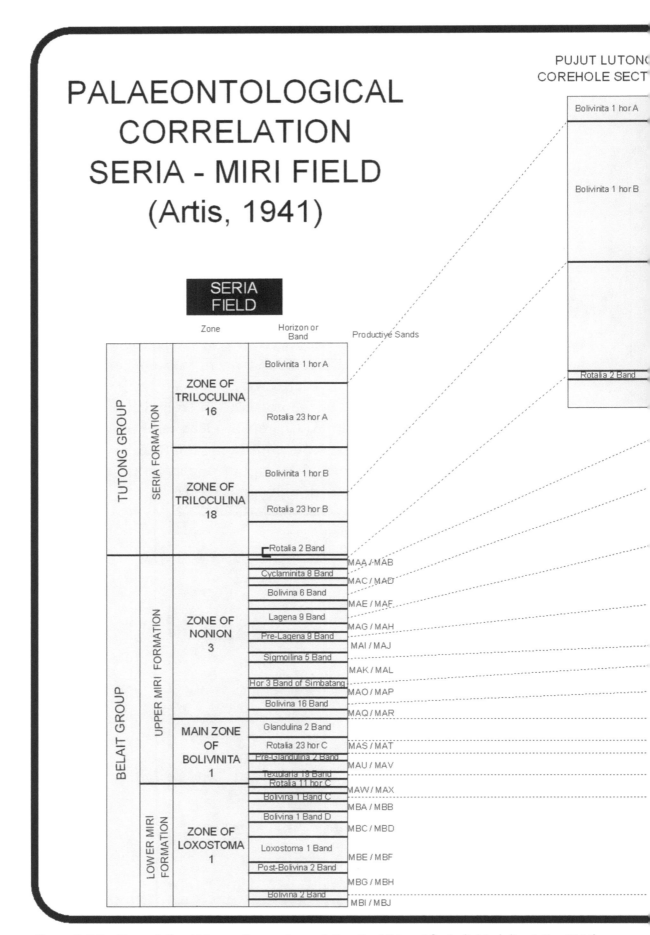

Figure 2.3.3a: Foraminiferal biozonation and correlation the Miri and Seria fields (after Artis, 1941)

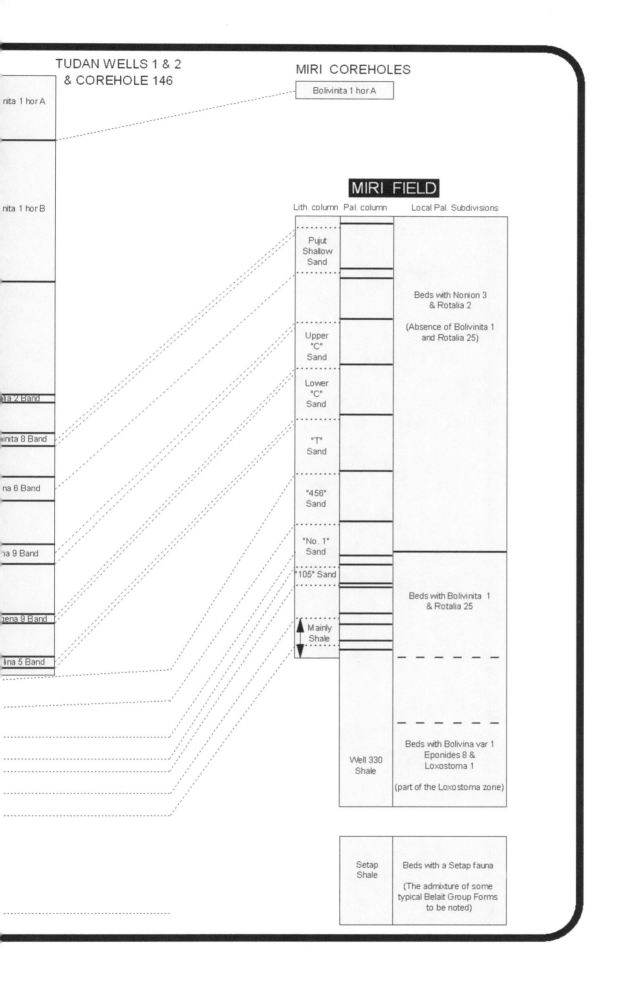

TUDAN WELLS 1 & 2
& COREHOLE 146

MIRI COREHOLES

Bolivinita 1 hor A

nita 1 hor A

nita 1 hor B

ita 2 Band

inita 8 Band

na 6 Band

na 9 Band

ena 9 Band

lina 5 Band

MIRI FIELD

Lith. column   Pal. column   Local Pal. Subdivisions

Pujut
Shallow
Sand

Upper
"C"
Sand

Lower
"C"
Sand

"T"
Sand

"456"
Sand

"No. 1"
Sand

"105" Sand

Mainly
Shale

Beds with Nonion 3
& Rotalia 2

(Absence of Bolivinita 1
and Rotalia 25)

Beds with Bolivinita 1
& Rotalia 25

Beds with Bolivina var 1
Eponides 8 &
Loxostoma 1

(part of the Loxostoma zone)

Well 330
Shale

Setap
Shale

Beds with a Setap fauna

(The admixture of some
typical Belait Group Forms
to be noted)

sand, with occasional calcareous streaks, and a large amount of incompletely carbonized plant remains, finely dispersed through the matrix. This formation is the main productive interval in the nearby Seria field in Brunei, where it reaches up to 5000 ft in thickness. Its age is poorly constrained by biostratigraphical methods; the Tutong outcrops are dated as Late Miocene (Late Tortonian to Early Messinian; biozone NN11A). The entire age range of the Serial Formation is likely to be Late Miocene (10-5 Ma). Based on foraminiferal faunas, Shell geologists recognized the presence of the Seria Formation along the southeastern flank of Canada Hill, where it is folded and stands up almost vertically. Thus, the thrusting that created the Miri structure postdates the deposition of the Seria Formation.

The Seria Formation is overlain by a marine transgressive succession of sands and clays with some lignites, of very young appearance and poor consolidation, overlain by similar fluvial, lagoonal or deltaic sediments which show considerable variations in thickness and lithology (Liechti et al., 1960), referred to as the Liang Formation. This formation is restricted to a narrow coastal belt mainly in Brunei, but is likely to extend as a thin veneer southward to the Bakam-Beraya area (see chapter 3.14). The Liang Formation is thought to be Pliocene in age (5-2 Ma).

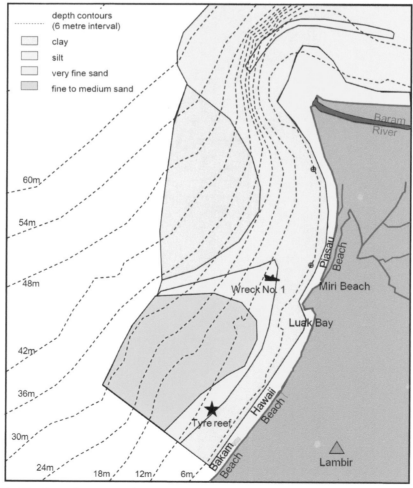

Figure 2.3.4a: Distribution of present-day sediments on the seabed west of Miri (after Lalanne de Haut et al., 1968). The offshore sands (yellow/orange) were dredged for the construction of Miri Marina.

Bordering Miri and extending offshore towards Brunei, a large delta system developed during the Pliocene: this (paleo) Baram Delta accumulated some 6 kilometers of fine-grained sediments over a period of three million years.

During the last few million years, a series of geological events took place that finally shaped the present-day configuration of northwest Sarawak. Compressional pulses along the crustal suture separating the South China Sea and Borneo produced folds and reverse faults along the margin of Borneo; continued uplift affected more prominently the inboard areas of Sarawak.

The present-day coastal area down to the West Baram Line suffered a phase of compression that created Lambir Hills, Miri Hill and the Seria structure. The Liang Formation was probably deposited at the same time as these structures were formed. The presence of fluvial conglomerates indicates that reliefs were actively eroded as the Liang Formation was being deposited. While the margin of northwest Borneo experienced some uplift, more significant vertical elevations affected the inboard areas: it is estimated that the Mulu area was uplifted by as much as 5 km over the past 17 million years. In the last two million years or so, ongoing uplift and erosion at Mulu created the cave systems that characterizes the Gunung Mulu National Park.

2.3.4 The Holocene (120,000y ago to present day)

Sarawak is bordered by a wide shallow sea (the Sunda shelf), which has repeatedly emerged and submerged during the various Holocene glaciations and interglaciations. In the latest glaciation (Weichselian), the global sea level was about 150 m lower than today. At the end of the Weichselian, climate warmed up, melting the extensive glaciers at higher latitudes, and causing a worldwide late Holocene sea-level rise and drowning of coastal environments.

Several kilometers offshore Sarawak, a broad band of sandy sediments has been mapped on the sea floor from southwest Miri to at least Jerudong in Brunei (Lalanne de Haut et al., 1968, Figure 2.3.4a). While these sediments have not been properly dated, their age may be between 8000y BP to the present. They are characterized by a rich mollusk fauna; many of these species have not yet been identified as living in the area today. These sediments were likely laid down during the initial Holocene sea-level rise. When the Baram delta started building out again (after 6000y BP), it partially covered these older sediments. In other areas, notably at two known dive sites (an artificial reef southwest of Tanjong Lobang and at Miri wreck No1), there has been little or no sedimentation since then, causing the older sands to be exposed at the seabed. At these sites, the sands contain the same fauna, but it is not clear when sedimentation ceased. The sands from the offshore layer have been dredged and used to construct the Miri Marina, where many fossils have been collected (see chapter 3.11.3).

At the high stand sea level, some 6000 years BP, the entire delta plains of the Baram, Miri, and Belait rivers drowned, forming a huge embayment bordered by the

Lambir and Belait hills (Figure 6 in Caline & Huong, 1992). Fossil-rich sediments found near Canada Hill indicate that the river mouths lay temporarily so far inland that a coral reef could formed behind the hill. These sediments comprise a varied fauna of corals and seashells (see chapter 3.11.1), including many species that presently are not known in the Miri area, and with their nearest occurrence at Labuan or Sabah. Sediments from an outcrop in front of Canada Hill (Sungai Baong) show that this area was exposed to the open sea. Here, the fauna comprises many species presently living in shallow waters surrounding Miri, but there are a lot more coral fragments and coral-related mollusk species that can be found on the beaches of Miri. These lines of evidence indicate that there was no fringing reef (of the type for example presently bordering Labuan Island), but smaller coral banks formed on the open shelf – where there is now too much sediment in suspension for corals to grow. Such coral banks currently only occur south-west of Tanjong Lobang.

In the course of time, the Baram, Miri and Belait rivers deposited so much clastic sediments that all reefs in the Baram embayment were quickly drowned by mud, whilst the water became brackish. Eventually, as the sediment supply continued, coastal barriers developed all along the northern Sarawak, closing off the embayment except for the outlets of the rivers.

Initially there were tidal flats behind these coastal barriers, as mentioned by Caline and Huong (1992, their figures 8-10) and evidenced from specific tidal flat mollusks washing up on the beaches south of Miri (where their preservation indicates a Holocene age). As the sand supply continued, the Baram river delta gradually took on its present shape.

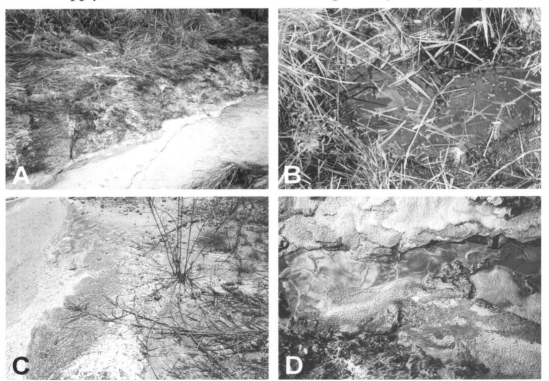

Figure 2.4a: Naturally outcropping oil seeps around Miri. A: outcrops along Jalan Padang Kerbau, Miri Formation; B: ditch on top of Canada Hill, near Well#1, Miri Formation; C: outcrops in Lambir Hills, Km 28 Miri-Bintulu Road, Lambir Formation; D: outcrops along Miri-Bekenu coastal road, Km 18.7, Miri Formation

2.4 Petroleum Geology of the Miri Oil Field

Hydrocarbons (liquid oil and natural gas) are generated from organic-rich sediments which have been heated to 70-120°C, at which temperatures the carbon-rich organic compounds (kerogen) start breaking down to simpler hydrocarbons. Being lighter than water, oil then migrates upward, from the source rock (usually black shale) into porous and permeable reservoir rocks (typically sandstones but also limestones), eventually capped by an impermeable seal rock (oftentimes claystones). Large oil accumulations in traps will then form commercial targets for exploration and production.

Active oil seeps found in and around Miri (Figure 2.4a) indicate that at least one oil source-rock has been expelling its product, or that oil in an underground reservoir is not completely trapped and a portion of it is leaking to the surface, probably along faults. Petroleum geologists study source rocks for their potential hydrocarbon generation, they examine reservoir rocks for their storage and yield capacity and their position with respect to sealing rock; they also try to determine the timing of hydrocarbon migration and formation of traps in the basin. Identifying the sites of hydrocarbon accumulations is the ultimate target of petroleum geologists, and this is not an easy job because traps are buried thousands of feet underground and one is never certain of an oil or gas accumulation until it is drilled. Structural deformations generate suitable petroleum traps which may be anticlines, fault blocks, or often a combination of the two, as is the case in the Miri oil field. Initially it was thought that the Miri structure consisted of a rather simple dome-shaped structure and the importance of faulting was not recognized. However, after drilling a large number of wells and studying many fault outcrops in Miri, it became understood that the Miri field was compartmentalized into numerous fault blocks. The geological map of the Miri field by von Schumacher (1941; Figure 2.4b) and further data and observations have corroborated the principles of this structural model.

Hills spanning from Tanjung Lobang to Pujut are the surface expression of the faulted Miri anticlinal structure, which extends further under the sea in a westward direction and plunges like a nose towards Tudan in a northeast direction. Canada Hill, or the eastern part of the Miri oil field, is bounded on both sides by reverse faults; it was home to the celebrated well Miri#1 as well as the great majority of wells drilled during the first half of the twentieth century in Miri. The Miri anticline is cored by diapirs of the Setap shale, penetrated in at least six deep wells where this formation is found to be overpressured and highly fractured.

The northwestern flank of the Miri structure is dissected by a conjugate set of NW-dipping and SE-dipping normal faults; virtually the entire oil production from the Miri field came from the fault traps located on the northwestern flank of the structure. The Shell Hill Fault and the Canada Hill Thrust delineate the upthrown central block of the Miri structure that follows the axis of the hill from Tanjung Lobang to Pujut; minor production came from this block. The Canada Hill Thrust in part overlies the southeastern flank of the Miri structure; this flank consists of subvertically folded clastics of the Seria Formation, with no production.

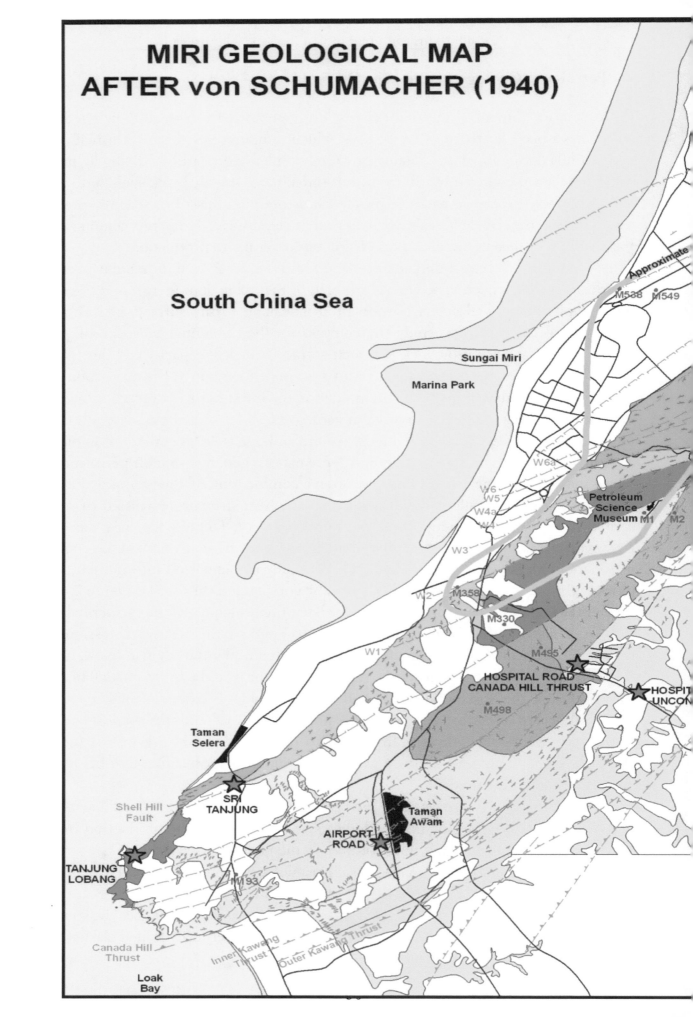

MIRI GEOLOGICAL MAP
AFTER von SCHUMACHER (1940)

South China Sea

Sungai Miri

Marina Park

Approximate

M538 M549

W6a

W6
W5

W4a

W4

Petroleum
Science
Museum M1   M2

W3

W2   M358

M330

W1

M495

HOSPITAL ROAD
CANADA HILL THRUST

HOSPIT
UNCON

M498

Taman
Selera

Shell Hill
Fault

SRI
TANJUNG

Taman
Awam

AIRPORT
ROAD

TANJUNG
LOBANG

M193

Canada Hill
Thrust

Inner Kawang
Thrust   Outer Kawang Thrust

Loak
Bay

Figure 2.4b: Geological map of the Miri Field
redrawn after von Schumacher (1941)

The structural model of the Miri field is hampered by lack of seismic images and poor resolution of stratigraphic data. The Shell Hill Fault is traditionally interpreted as a major seaward-dipping growth fault; the delineation of this fault is based on surface expression (where present) and on subsurface correlation of various sandstone members of the Miri Formation, which is a formidable task given the poor quality of well samples and the overall low resolution of biostratigraphic data. Fault aliasing and misinterpretations are highly probable, as exemplified in Schumacher's geological map where the Shell Hill Fault obliquely crosses the main structure (Canada Hill). Wells drilled on both sides of the Shell Hill Fault show a fairly constant thickness of the sand units and there is no systematic evidence for growth of the sedimentary units across the fault (Shuib, 2001). In the Sri Tanjung outcrop, the Shell Hill Fault juxtaposes the "Upper C Sands" against the "Lower C Sands" (see chapter 3.3), with a displacement possibly in the order of tens of meters. If, as is shown on Schumacher's geological map (Figure 2.4b), the Shell Hill Fault places the "Pujut Shallow Sand" against the "105 Sands", then the growth of the fault may reach many hundreds of meters.

According to the traditional interpretation, the southeastern flank of the Miri structure consists of double reverse faults, with the Canada Hill Thrust and the Kawang Thrust representing respectively the inner and outer thrusts. While outcrop and well data clearly demonstrate the presence of a thrust zone along the southeastern extent of the hill, it is not proven that the thrust zone is partitioned into several thrust fronts. The thrust zone identified at the base of the steep southeastern hill front and the narrow and high topographical relief that characterizes the central and northern part of Canada Hill strongly suggests a pop-up structure, implying a second, north-western thrust front along the seaward side of Canada Hill. Several SE-dipping fault planes have been identified along Pujut Road by Shell geologists, and interpreted as normal faults; these steeply dipping faults may as well be reverse faults. Right-lateral transpressional movements parallel to the axis of the hill provide a likely explanation for the formation of the Miri structure, which is possibly a pop-up (positive flower) structure (Shuib, 2001).

It must be remembered that fault interpretations in the late 1930s were largely unconstrained by conceptual models and analogs. Given the poor stratigraphic resolution and questionable correlation of sandstone units in various parts of the field, a re-examination of the Miri structure is a task ahead of us. A rare seismic line acquired across the axis of the Miri structure to the north of the field (Rijks, 1981; Figure 2.4c) clearly shows that both the southeastern and northwestern flanks of the structure correspond to a reverse fault. It is noteworthy that none of the secondary eastern and western extensional faults are apparent on this seismic image. Morley et al. (2003) have argued for the reactivation (inversion) of older extensional faults as younger reverse faults.

The geological map of Miri, published in a Shell internal report in 1941, and reproduced by PETRONAS (1999, p.303), still represents our best effort today (Figure 2.4b). Nevertheless, the reliability of the mapped sandstone bodies and the fault systems still remains poor, and with better data the geological map of Miri will be improved over time.

Despite a long history of field production, there is still an open debate as to what source rock charged the Miri oilfield, because no mature source rock was penetrated by the wells. Given the lush tropical forests that have lined up along the sea shores for millions of years, it is likely that washed out land-plant rich in organic matter accumulated in subsurface stratigraphic intervals. Alternatively, the oil may have come from sedimentary beds rich in marine-derived organic matter, such as algae and other microorganisms. The Setap Shale may also be viewed as the most probable source rock for the Miri field.

Our inability to pin-point a specific source rock in the subsurface implies that we do not know when oil generation in the field started. A deep source rock would reach thermal maturity earlier than a shallow source rock. Because the structural traps in the field formed only a few million years ago, oil is probably still being generated as it is leaking to the surface today. We may thus assume that oil generation also started only a few million years ago. It should further be noted that while commercial volumes of oil have been recovered from the Miri field, significant quantities of associated gas have also been encountered.

Unlike the source rock, identification of the reservoir and seal rocks has been quite easy, since many sandstone beds with good reservoir quality and shale beds with good sealing capacity have been mapped in the outcrops around Miri. Indeed, seven productive sandstone units in the Miri Formation, each capped by shale have been identified by drilling numerous wells in the Miri field (Figure 2.4d). The productive sandstone units (payzones) were named haphazardly as they were discovered, such that the reservoir nomenclature appears disparate. The sealing shales proved to be characterized by different associations of microfossils (foraminifera); their nomenclature reflects the name of the key species that defines the shale unit.

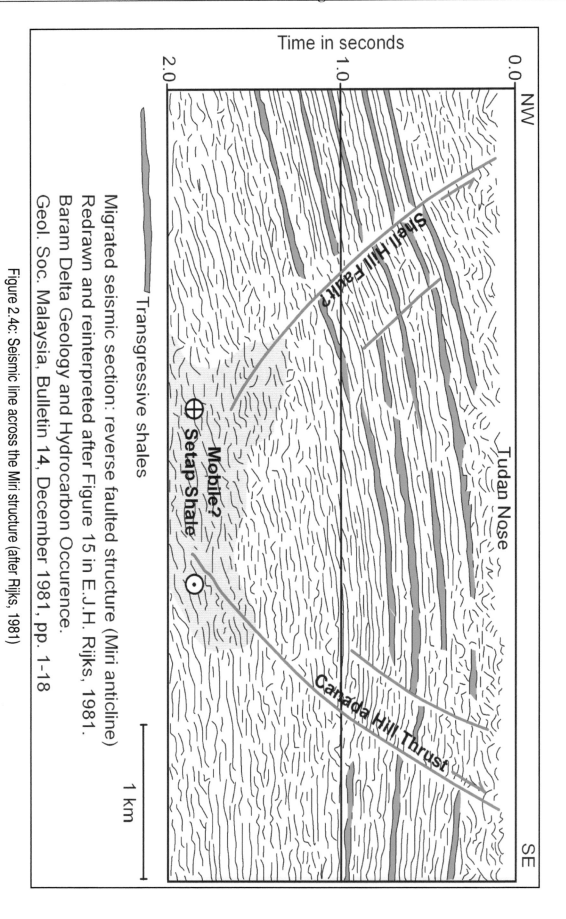

Figure 2.4c: Seismic line across the Miri structure (after Rijks, 1981)

Migrated seismic section: reverse faulted structure (Miri anticline)
Redrawn and reinterpreted after Figure 15 in E.J.H. Rijks, 1981.
Baram Delta Geology and Hydrocarbon Occurence.
Geol. Soc. Malaysia, Bulletin 14, December 1981, pp. 1-18

Transgressive shales

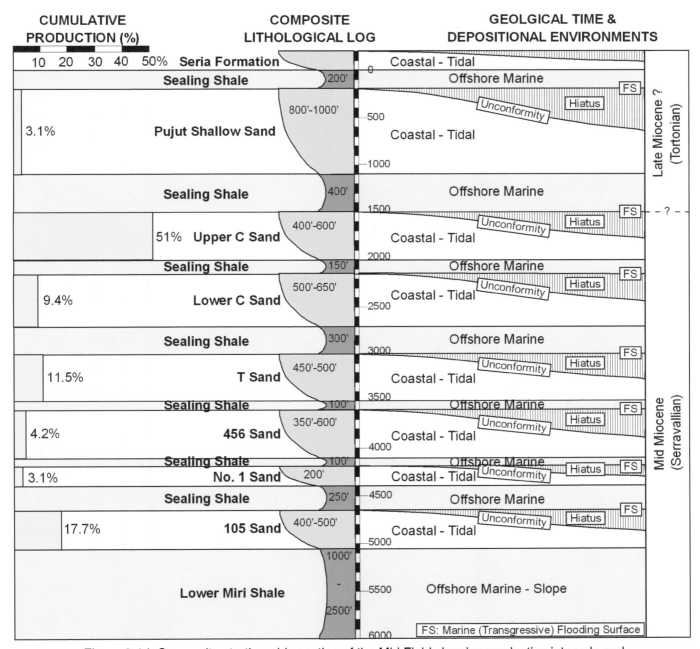

Figure 2.4d: Composite stratigraphic section of the Miri Field showing productive intervals and depositional environments (after von Schumacher, 1941)

Chapter 3

Geological Excursions

3.1 Foreword

As is customary in tropical areas, outcrops are ephemeral: they crumble because of erosion (Figure 3.1a), they become overgrown by vegetation or they disappear because of construction work. The geological excursions proposed here are based on fieldwork carried out by the authors from 1994-2010; most of these outcrops have been accessible for many years and should hopefully remain accessible in the foreseeable future. However, urban encroachment in and around Miri is likely to render access to some outcrops more difficult with time.

New outcrops are continuously being exposed, that may offer opportunities to study the sedimentary rocks, observe fault structures, collect fossils, or witness active hydrocarbon seepages.

Wherever an excavation takes place, outcrops will be exposed and conditions for observations are optimal. When studying outcrops, you will need a hand lens and a geological compass; take notes, as much as possible, of the following points:

- General description of the outcrop: Are the sediments continuous all over the outcrop, or are there specific changes vertically or laterally?
- Look for a representative part of the outcrop to measure a stratigraphic section: using a measuring tape, start at the bottom and describe every layer, taking note of sediment type, color, presence of structures (sedimentary or structural), fossils, etc. The reference point could be ground level or average sealevel.
- Make a sketch of the outcrop in your notebook. Indicate the location of your pictures, of the measured section, fossils and any other geologic features of interest. If you have a GPS, note the coordinates.
- If fossils are collected, label them properly. It is important not to mix fossils coming from different beds within a same outcrop. On the label indicate the location of the outcrop, stratigraphic position (bed level), name of collector and date of collection, and (if known) the scientific name of the fossil(s).
- Back home, redraw your log at scale, on millimeter paper, making use of a standard legend such as the one used for stratigraphic logs in this guidebook (Figure 3.1b). On your computer, create a workarea for each outcrop where you can store all relevant data and pictures.

If you have never studied an outcrop before, but have discovered an interesting exposure, you may seek assistance from a professional or amateur geologist. This may be achieved through the local geological centers, or by contacting one of the authors of this guide.

We have listed the safety hazards particular to each outcrop. Be very cautious however, because conditions change all the time and new hazards may develop. Always remember to give priority to safety, and follow some simple rules:

- If you drive and see an outcrop of interest, keep your eyes on the road. For better examination, park your car in a safe location and walk to the outcrop
- Stay away from steep cliffs and other hazardous spots

- Ask for permission before entering active quarries, excavation sites or private lands
- Minimize environmental impact (damage to the outcrops as well as living plants and animals)
- Be equipped with a hat, shoes that will protect your ankles, sun block cream, and plenty of water
- Have a cellphone with you
- Use common sense

Figure 3.1a: Tanjung Tusan cliffs before and after their collapse in the spring of 1999

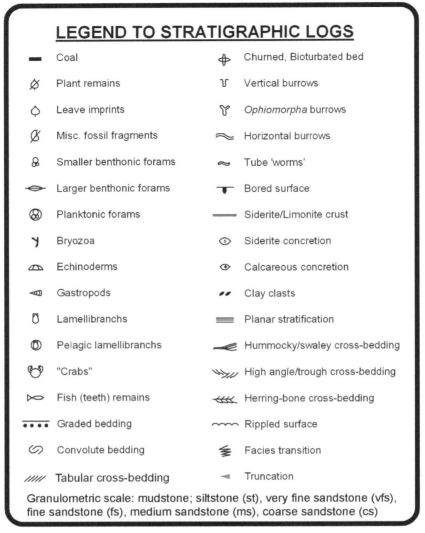

LEGEND TO STRATIGRAPHIC LOGS

▬	Coal	⊕	Churned, Bioturbated bed
⌀	Plant remains	Ỻ	Vertical burrows
◌	Leave imprints	Ψ	*Ophiomorpha* burrows
⌀	Misc. fossil fragments	≈	Horizontal burrows
8	Smaller benthonic forams	≈	Tube 'worms'
⬬	Larger benthonic forams	⊤	Bored surface
⊗	Planktonic forams	▬▬	Siderite/Limonite crust
Ƴ	Bryozoa	◉	Siderite concretion
⌒	Echinoderms	◉	Calcareous concretion
⬦	Gastropods	‚‚	Clay clasts
◖	Lamellibranchs	≡	Planar stratification
◎	Pelagic lamellibranchs	◢	Hummocky/swaley cross-bedding
⬮	"Crabs"	⤳	High angle/trough cross-bedding
▷	Fish (teeth) remains	⋘	Herring-bone cross-bedding
▰▰▰	Graded bedding	⌇	Rippled surface
⟲	Convolute bedding	⦚	Facies transition
/////	Tabular cross-bedding	◀	Truncation

Granulometric scale: mudstone; siltstone (st), very fine sandstone (vfs), fine sandstone (fs), medium sandstone (ms), coarse sandstone (cs)

Figure 3.1b: Legend to the stratigraphic logs

3.2 Tanjung Lobang

3.2.1 Access and Location

From the Clock Tower roundabout at km 0.0, make your way southeastward towards Taman Selera Park, via Jalan Kubu and Jalan Temenggong Oyong. On the way, you will be passing the Parkcity Everly Hotel and the Marriott Resort; Taman Selera is a recreation park facing the sea and has plenty of parking lots (N 04° 22.148'; E 113° 58.080', Figure 3.2.1a).

3.2.2 Logistics and Safety

The path from Taman Selera to the cliffs of Tanjung Lobang (Stop 1 to Stop 4) is the most demanding and dangerous area of field excursions described in this guide book (Figure 3.2.1a). You need to be fit, wear shoes with a good grip to follow the whole tour, and it is best to visit this outcrop in a group or accompanied by a friend. Parts of the tour are not well marked, and the path may be lost over some distances, as only the occasional hasher transits it. We strongly recommend that you bring your mobile phone with you, or that you inform a person and leave instructions regarding your field excursion. Plan your visit at low tide. To return to Taman Selera it is best to follow the walkway along Jalan Tanjung Lobang. The whole excursion will take half a day, so take plenty of water and some snacks.

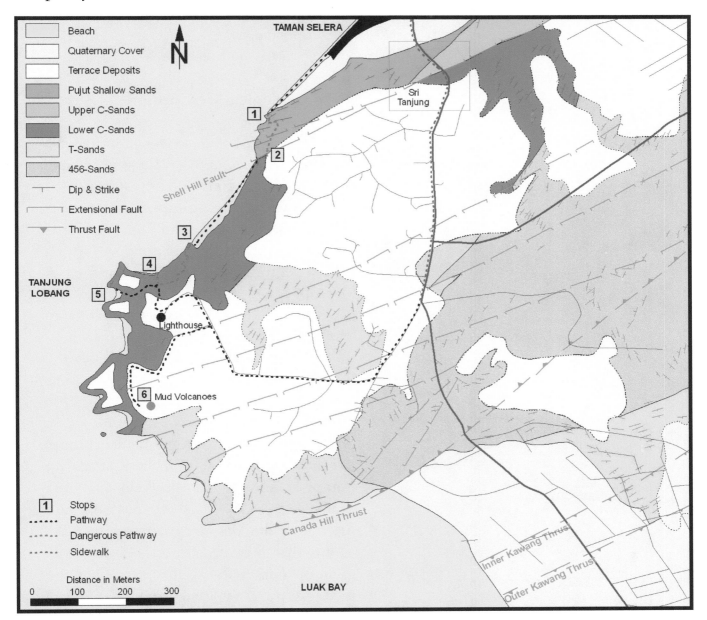

Figure 3.2.1a: Location and geological map of the Tanjung Lobang outcrop

WARNING: Do not proceed if the weather is rainy, or after much rain. Parts of the tour will take you onto slopes plunging into the sea, and to steep trails in the forest that can only be managed under dry conditions. Part of the trip will lead you close to high cliffs, which are unstable as shown by the many blocks scattered at the foot of the cliffs. Always remain at a safe distance from the cliffs.

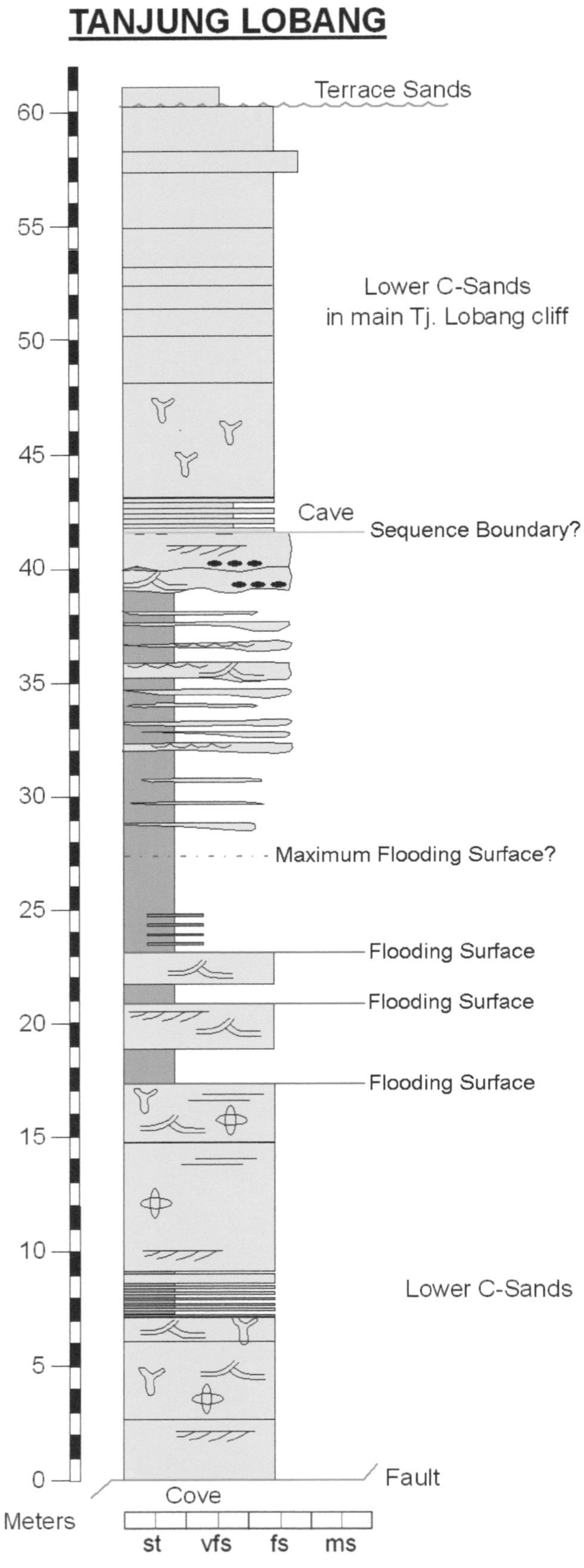

TANJUNG LOBANG

Figure 3.2.3a: Stratigraphic log of the "Upper C Sands", Tanjung Lobang outcrop

3.2.3 Outcrop Highlights

- The "Pujut Shallow Sands" and "Lower C Sands" (Figure 3.2.3a)
- Sequence stratigraphy of the "Lower C Sands"
- Expression of faults and terraces in the landscape
- Extinct mud volcanoes

3.2.4 Geological Description

The first part of the trip will get you to the base of the main cliff of Tanjung Lobang through some convoluted pathways and will give you an opportunity to observe the "Pujut Shallow Sands" and the "Lower C Sands" of the Miri Formation.

From your parking location at Taman Selera, follow the beach in the direction of the cliffs, readily visible from the park (Figure 3.2.4a). After some 5-10 minutes of walking, at the end of the beach (Stop 1), a sandstone sequence with larger concretions and rare *Ophiomorpha* burrows is encountered (see chapter 4.1.4). These gently seaward dipping sandstones are part of the "Pujut Shallow Sands", a minor reservoir in the Miri Field. Continue walking carefully on the sandstone, and note how the *Ophiomorpha* traces increase along this outcrop (Figure 3.2.4b). The outcrop comes to an abrupt end, as the beds are offset by a fault oriented perpendicular to the coast (Figure 3.2.4c).

Look for a path in the vegetated area a little before the fault zone, and climb to reach a ridge that marks an inland continuation of the fault. The path is poorly marked and climbs abruptly; it then follows the edge of the fault characterized by a steep cliff. This passage is dangerous and demands full attention; from the beach, some 15 minutes will be needed to reach the top of the cliff (Stop 2). The spot offers a great photo opportunity of Tanjung Lobang cliffs and a larger cove in the foreground.

Walk down towards the cove through the forest; this part is poorly marked and requires some agility to pass through the bushes and down the steep slope. The beach is reached in a few minutes, and the first visit should be the base of the cliff. Here, an oblique NE-SW trending fault can be observed, which you can follow through the forest; this is the Shell Hill Fault, extending from the Sri Tanjung outcrop (see Figure 3.2.1 and Chapter 3.3). Oil seeps are at times visible along the fault.

At the southwestern end of the beach (Stop 3), a NW-SE trending fault exposes a 20-m thick sequence of sandstones ("Lower C Sands"), which has produced a high, overhanging cliff with spectacular *Ficus* trees and hanging root systems.

At this point, the path follows the seaward dipping sandstone bed, and winds up into the forested area in a zig-zag fashion; this path may be dangerous at times and is always difficult to walk through. It soon reaches a height corresponding to that of the terraces above the cliffs of Tanjung Lobang where the path is again well marked. The path downward along a gully leads to the sea, at the base of the main Tanjung Lobang cliffs (Stop 4), which are delineated again by a NW-SE trending fault zone. Once on the shore, walk up a low ridge of sandstones to the right in order to gain an overview of the cliffs (Figure 3.2.4d):

Four different sedimentary units comprise the Tanjung Lobang cliffs:

1) Bedded sandstones at the very base of the cliff, extending from the previous cove
2) Grey-blue mudstones with thin sandstone beds, at the lower part of the cliff
3) Amalgamated yellowish sandstones, forming the main cliff
4) A thin unit of reddish and white soft sandstones, capping the cliff.

The first three units form part of the "Lower C Sands" of the Miri Formation. Closer examination of each of these units throws light on their depositional history.

Figure 3.2.4a: Tanjung Lobang cliffs from Taman Selera

Figure 3.2.4b: Seaward dipping sandstone with *Ophiomorpha* trace fossils,
"Pujut Shallow Sands", Tanjung Lobang outcrop

Figure 3.2.4c: Small fault offsetting the sandstone beds, Tanjung Lobang outcrop

Figure 3.2.4d: Tanjung Lobang cliffs overview and interpretative diagram.

At the far end of the beach, the lower sandstones are very fine-grained, laminated at the base and become massive with *Ophiomorpha* burrows at the top (Figure 3.2.4e). Presumably, these sandstones were originally deposited in a shallow sea where marine currents created the laminations; liquefaction of the sands and/or intense bioturbation obliterated the laminations and resulted in the massive appearance of the sandstones.

This location is most appropriate to take a measurement of the dip of the sandstone beds, which can be measured perpendicular to the horizon (strike) of the beds; the beds dip in a westerly direction with an angle of some 23°.

Unit 1 is not fully exposed; it consists of 1.5 m thick silty and sandy shales, passing upward into amalgamated sandstones some 2 m thick; these crop out a few meters away in the direction of the cliff. The section continues in mudstones and interbedded thin sandstones, passing upward into 1.5 m thick massive, bioturbated sandstones, with possible hummocky cross-stratification.

The stratigraphic succession up to this point is indicative of shallow marine conditions, possibly no deeper than 20 m, at which depth conditions are adequate for swells to mobilize the sediments and create hummocky cross-stratification. Sediment accumulation was keeping pace with the basin subsidence, resulting in an aggrading series of sandstones within the overall identical, shallow-marine depositional environment.

Figure 3.2.4e: Massive sandstone with *Ophiomorpha* trace fossils, "Lower C Sands", Tanjung Lobang outcrop

The continuation of the stratigraphic section marks a significant change in lithology, with the deposition of Unit 2, a 16 m thick unit of mudstones, exposed at the base of the cliff (Figure 3.2.4f). The depositional environment of these mudstones corresponds to a marine setting, far away from the coastal sandstones and at relatively greater water depths.

The lower half of this unit consists of silty mudstones, with cm-thick, reddish silty concretions; it is devoid of sandstones. The upper half is characterized by the presence of cm-dm thick sandstone beds, becoming more frequent towards the top, and ending up in an amalgamated interval.

The individual beds of sandstones show significant lateral thickness variation over the length of the outcrop, including lateral thinning out. The base of the sandstones is in part erosional, forming incised gutters that cut into the encasing mudstones (Figure 3.2.4g). The original seabed must have been scoured by a series of incisions forming elongated gullies that were created by strong local currents, probably storm events.

The sands filling these gullies are fine-grained, and show hummocky cross-stratification; clay clasts are common at the base of the sandstones. Spillover sandstones extend away from the gutters as 10-20 cm thick, laminated units, which become thinner laterally over a distance of 5-10 m. Shell lags are sometimes present at the base of the gutters, while ripple marks can be observed at the top of the sands and into the mudstones. The uppermost beds are 2-3 m thick, and consist of fine-grained, hummocky cross-stratified and amalgamated sandstones (Figure 3.2.4h), with multiple erosive surfaces and occasional thin, discontinuous mudstones intervals. The top of this bed is in sharp contact with the aggrading series of sandstones above, which form part of Unit 3.

Figure 3.2.4f: Marine mudstones with thin sandstones
indicating a marine transgression, "Lower C Sands",
Tanjung Lobang outcrop

In summary, this second sedimentary unit is composed of relatively deeper-water and distal offshore sediments. Three significant stratigraphic surfaces define this interval:

- The base of the mudstones is a flooding surface, recording an abrupt deepening of depositional environments
- A maximum flooding surface is probably located slightly below the base of the first, thin interbedded sandstone
- The top of the hummocky-stratified, amalgamated sandstones, in contact with the overlying thick amalgamated sandstones is a possible sequence boundary; it records a sharp return to much shallower depositional environments and a new trend in stratigraphic architecture, namely an aggrading pattern.

The third sedimentary unit consists of aggrading bedded sandstones, a relatively harder unit, with constitutes the Tanjung Lobang cliffs. This unit is about 15 m thick; it cannot be safely accessed, and could therefore not be logged in detail. From a distance, the sandstones appear to be current-bedded and represent a return to a coastal, shoreface environment. *Ophiomorpha* burrows can be recognized from afar because their limonitic outlines in sandy units are so bioturbated that the original bedding has been obliterated (Figure 3.2.4i). Some fallen blocks have yielded uncommonly large burrows, with a diameter of some 10 cm.

Figure 3.2.4g: Erosional base of a sandstone unit embedded within marine shales, "Lower C Sands", Tanjung Lobang outcrop

Through wave actions, caves are currently forming at the base of the sandstones, as they plunge into the sea; this erosional process progressively undermines the stability of the cliffs, and will eventually lead to their collapse.

The uppermost sedimentary Unit 4 consists of a thin veneer of friable, sandy sediments, which unconformably overlies Unit 3. It is approximately a meter thick horizontal bed, forming the base of a terrace that truncates the dipping Miri Formation sandstones. The base of the unit is a reddish, dirty sandstone; it is overlain by leached, well-sorted, very fine-grained sandstone. The Terrace Unconformity at the base of Unit 4 is a major sequence boundary, with a time-gap of about 10 million years.

The excursion continues by retracing the steps back to the top of the terrace; past the main cliff, a path leads down to a small cove situated behind the previous section (Stop 5). The path follows fault lines and exposes some of the mudstones; on the way, there are spot occurrences of oil-impregnated sandstones.

This short excursion to the beach area provides an opportunity to look at some stacked shoreface sandstones, not previously accessible. Internal scour surfaces can be recognized at various levels; some sandstones display concentric, red-colored rings (the so called "Liesegang Rings") that are diagenetic effects resulting from the diffusion of iron-bearing solutions within the sandstones (Figure 3.2.4j).

While returning to the top of the terrace, the uppermost reddish and white sands (Unit 4) can again be observed, in unconformable

Figure 3.2.4h: Amalgamated, hummocky cross-stratified sandstones, "Lower C Sands", Tanjung Lobang outcrop

with the underlying Lower C Sands. From the top of the terrace, a path continues to the Light House and the adjacent Japanese War Memorial. From here, a pathway leads out of the forest, passing by some houses and reaching the college Tun Datu Tuanku Haji Bujang. Past a carved gate on the right, a concrete path leads you back to the edge of the cliffs, following a well-marked terrace. This 10-min walk ends in the southernmost Tanjung Lobang cliffs (Stop 6) and affords spectacular views of the last cove and towards the Lambir Mountains. An isolated block of sandstones stands out of the sea at a distance from the cliffs, indicating the significant erosion suffered by the cliffs.

Figure 3.2.4i: Massive sandstone with *Ophiomorpha* trace fossils, "Lower C Sands", Tanjung Lobang cliff

Figure 3.2.4j: Iron enrichment zones as Liesegang Rings, "Lower C Sands", Tanjung Lobang outcrop

Figure 3.2.4l: Tanjung Lobang Cliff terminating at a major fault zone

Close to the end of the path, a series of inactive mud volcanoes can be observed some ten meters inland (Figure 3.2.4k). The conical mounds are about 1.5 m high, and one of them has its crater morphology well preserved. These mud volcanoes are the surface expressions of a leaky, over-pressured shale system that has brought fluidized muds and gases to the surface, along fault and fractures.

The abrupt termination of the cliffs, visible from a short distance (Figure 3.2.4l), is due to a reactivated fault. The excursion finishes at this point. Walk back to the Lighthouse road, crossing the grounds of the college linking with Jalan Tanjung Lobang, and following the sidewalk down the hill back to Taman Selera. This return walk takes about half an hour and should be linked with the Sri Tanjung outcrop observation, described hereafter.

Figure 3.2.4k: Remnant cone of an inactive mud volcano, Tanjung Lobang outcrop

3.3 Sri Tanjung

3.3.1 Access and Location

From the Clock Tower roundabout at km 0.0, make your way southeastward (towards the Tanjung Lobang cliffs), via Jalan Kubu, and Jalan Temenggong Oyong, leading to Jalan Tanjung Lobang. On the way, you will be passing by Parkcity Everly Hotel, Marriott Resort, and Taman Selera. You may park at Taman Selera and walk up the hill, or drive up the hill and turn left to the first side road upon reaching the hill top. Park under the casuarina trees leading to the entrance of Rumah Kediaman Gunasama Persekutuan (N 04° 21.970'; E 113° 58.184', Figure 3.3.1a).

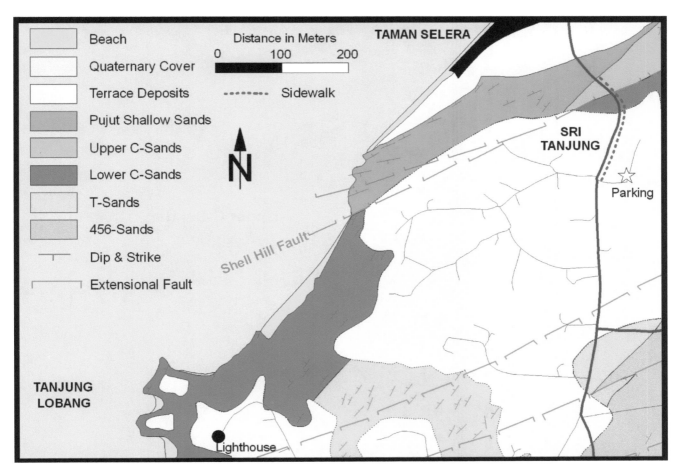

Figure 3.3.1a: Location and geological map of the Sri Tanjung outcrop

3.3.2 Logistics and Safety

This visit involves a 20-30 minutes walk on the sidewalk of Jalan Tanjung Lobang. Remain on the sidewalk, do not cross the road and be vigilant with the traffic. An umbrella might be useful to protect you from strong sunshine. The morning light is particularly well suited to view the outcrop.

3.3.3 Outcrop Highlights

- The western flank of the Miri anticline.
- The Shell Hill Fault with a water spring along the fault contact.
- Exposure of the "Upper C Sands" that accounted for almost half of the Miri field production (Figure 3.3.3a).
- The Terrace Unconformity

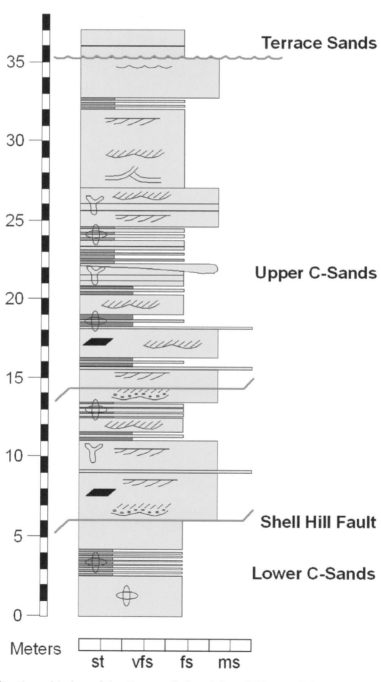

Figure 3.3.3a: Stratigraphic log of the "Lower C Sands" and "Upper C Sands", Sri Tanjung outcrop

3.3.4 Geological Description

From your parking spot on the hilltop walk down Jalan Tanjung Lobang in the direction of Taman Selera, following the sidewalk. Halfway down the road, as the road bends, the Miri Formation is exposed on the opposite side of the road, behind a stone wall; this forms part of the Sri Tanjung residence and cannot be accessed. At the time of writing, the outcrop was fresh and free from vegetation cover, allowing good observation from a distance.

Walking down the hill, the stratigraphic succession of the "Lower C Sands" (Figure 3.3.4a) is interrupted by a major normal fault, identifiable by a sharp break in the succession of the beds, by a change in dip direction, by a noticeable drag of sandstone beds on the right-hand side of the fault zone, and by the presence of a spring at the base of the fault (vegetation above the wall, on left-hand side; Figure 3.3.4b). As the fault plane crosses the road, the percolating water undermines the stability of the pavement and creates holes that demand frequent repairs.

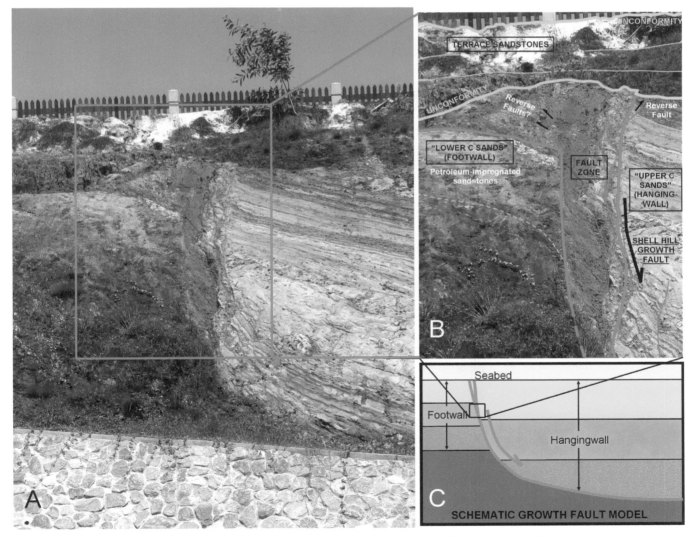

Figure 3.3.4a: (A) Detailed view of Shell Hill Fault and the Terrace Unconformity, Sri Tanjung outcrop; (B) Interpretative diagram of the Shell Hill Fault; (C) Schematic model of a growth fault

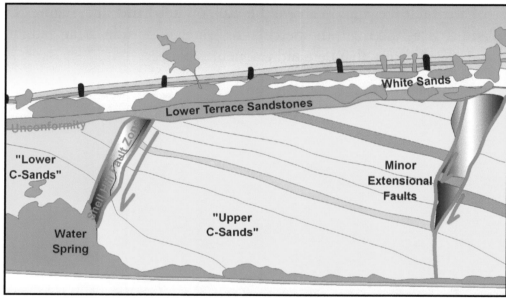

Figure 3.3.4b: Oblique view of the Shell Hill Fault truncated by the Terrace
Unconformity: Picture and interpretative diagram, Sri Tanjung outcrop

This is the Shell Hill Fault, which seals the petroleum-impregnated sandstones of the "Lower C Sands" against a complex fault zone, having experienced extensional growth, before being reactivated as a minor reverse fault. Vertical thickening of the "Upper C Sands" can be observed next to the Shell Hill Fault (Figure 3.3.4a A,B) and further away, down the hill (Figure 3.3.4c); this lateral increase in the thickness of the hangingwall block is characteristic for syn-sedimentary growth along the fault plane (Growth Fault: see Figure 3.3.4a C). Evidence for inverse reactivation of the Shell Hill Fault can be observed along the upper part of the fault zone, below its truncation by the Terrace Unconformity (Figure 3.3.4a B), where some smaller slivers of the "Lower C Sands" are clearly up-thrusted. This probably reflects a surface stress-release of the fault zone and is not related to a tectonic phase; the surface bulge of the fault zone is another expression of late stress-release. The apparent relief shown by the Terrace Unconformity above the fault zone is an indication that most of the surface stress-release of the fault zone happened during the phase of erosion that preceded the deposition of terrace deposits.

The throw along the Shell Hill Fault cannot be measured in this outcrop; however, as the fault appears to juxtapose two different formations against each other, the displacement is likely to be in the order of at least tens of meters.

The well-bedded "Upper C Sands" form a steeply plunging flank in the direction of the beach (Figure 3.3.4c). Their top is sharply truncated by terrace deposits, characterized by leached, white sands resting on dark brown, peaty sands. This angular Terrace Unconformity can be traced uphill, where it truncates the Shell Hill Fault and the "Lower C Sands" (Figure 3.3.4a).

The outcropping "Upper C Sands" are quite sand rich, with a net-to-gross sand ratio of 88%. The lower 19 meters of this unit consist of thicker sandstone beds with frequent occurrences of lignite streaks and lignite clasts, separated by heterolithic and densely burrowed sandy mudstones. The sandstones are dominantly current bedded; the coarser sandstone beds have a distinct yellowish colour, and show large cross-beds. Downcutting and laterally thinning sandstone beds correspond to channel deposits. The upper part of the outcrop is characterized by stacked sandstone beds with waveform bedding, including hummocky cross stratification.

The "Upper C Sands" correspond to a high-energy, coastal depositional environment with some tidal influence in the lower part, and storm sands in the upper part.

Figure 3.3.4c: An overview of the lower part of the Sri Tanjung outcrop, where vertical growth can be observed within the "Upper C Sands" (note stone wall along roadside).

71

3.4 Airport Road Outcrop, Heritage Site

3.4.1 Access and Location

From the Clock Tower roundabout at km 0.0, follow Jalan Miri Bintulu southward (towards the airport). After 2.5 km, as the road winds uphill, turn right into Jalan Medical Store; shortly after this bifurcation, take the first road on your left, and drive further uphill for a couple of hundred meters and park in the vicinity of a geological poster (No.1) on your right (N 04° 21.813'; E 113° 58.735'). From the parking area, walk downhill about 10 meters along the road, until you reach a small cross road on your right; here enter a small pathway on your left. As you enter the outcrop, a little to your left (approx.10 o'clock) you will see geological poster 2 in the distance.Cross part of the outcrop going downward towards your right, and use the handrails for added satefy until you reach Jalan Miri Bintulu. To start the visit, walk uphill along the pathway parallel to Jalan Miri Bintulu, until you reach the entrance of the quarry on your left and locate the geological Poster 3.

To access Taman Awam, drive 3 km from the Clock Tower before entering the Taman Awam parking area, on the left. Drive down to the far end of the parking area and stop. The Airport Road outcrops, part of an abandoned quarry, are located on the opposite side of the Jalan Miri Bintulu (N 04° 21.879'; E 113° 58.764', Figure 3.4.1a).

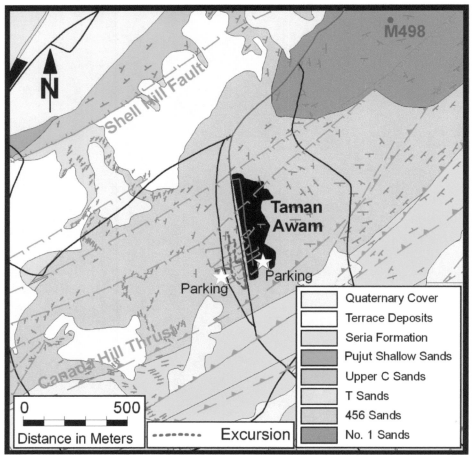

Figure 3.4.1a: Location and geological map of Miri Airport Road outcrop (M498: location of well Miri#498)

3.4.2 Logistics and Safety

Depending on your interest in faults, this 1-2 hours stop can take the whole day! The visit consists of a tour of the old quarry along a pathway the geology of which is described in 3 posters installed at the site. This outcrop is Sarawak's first protected geological site, under the auspices of the Miri City Council (see also Figure 5.4.1a and Chapter 5).

In the quarry, keep a safe distance from the steep faces; wearing of helmets is recommended for people wanting to study the lower cliff face in detail.

The outcrop has been described by Burhannudinnur & Morley (1997), Lesslar & Wannier (2001), van der Zee (2001), and Sorkhabi & Hasegawa (2005).

3.4.3 Outcrop Highlights

- The southern apex of the Miri anticline
- A series of normal faults that can be viewed in 3-dimensions over some 100 meters
- Exposure of the "456 Sands", a minor producing reservoir of the Miri field (Figure 3.4.3a)
- Wave and current induced sedimentary structures including text-book examples of herring bone cross bedding

3.4.4 Geological Description

Start the visit at Poster 3 standing in the middle of the quarry floor and facing a series of distinct normal faults (Figure 3.4.4a). The poster gives explanations of the structural geology of the outcrop including a brief introduction to fault analysis.

In this outcrop, the

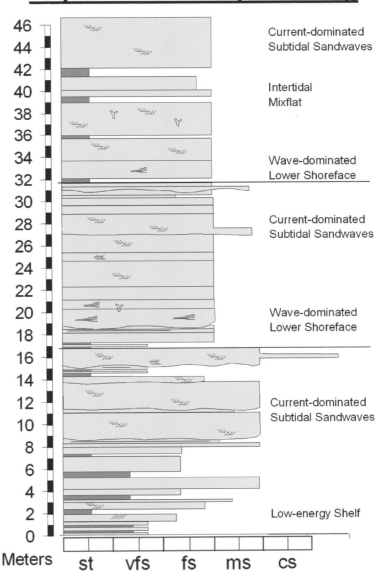

Figure 3.4.3a: Stratigraphic log of the "456 Sands", Miri Airport Road outcrop

exposed "456 Sands" of the Miri Formation have an overall high net-to-gross sand content (around 85% sandstone) and consist of a series of coarsening upward cycles. Each cycle starts with mud-rich heterolithics, and passes upward into mud-prone sandstones with wave-generated sedimentary structures; cleaner and often coarser sandstones with current-generated sedimentary structures cap the cycle of sediments. Each cycle starts with relatively distal and deeper water sediments (wave-generated facies), and becomes progressively proximal with shallower water sediments (current-generated facies) at the top, reflecting an overall progradation.

Some 20 normal faults can be observed throughout the quarry outcrop, which have been labeled according to their location (Figure 3.4.4b). The segment facing the poster exposes the most spectacular faults, labeled C1 to C8, from left to right. The opposing wall of the quarry to the left of the poster exposes faults B1 and B2, while faults A1 to A10 are located on the lower side of the quarry, in the direction of the airport.

The ground at the base of Poster 1 exposes the top of a sandstone bed, characterized by a cross-cutting joint pattern; these tension cracks were formed as a response to the folding of the rocks. Joints form (propagate) in the direction parallel to the maximum stress; in our case, the main joints extend at an angle of 280° while the subordinate joints have an angle of 325°. This direction is perpendicular to the axis of the Miri

Figure 3.4.4a: Stratigraphic log of the "456 Sands", Miri Airport Road outcrop

anticline and may indicate a late, stress-releasing mechanism, as the rocks were folded.

From left to right, 8 faults can be observed on the western quarry wall, referred to as C1 through C8. All faults are extensional and dissect centimeter to meter bedded series of sandstones and heterolithics. Distinctly colored marker beds allow a correlation of sedimentary layers from fault to fault. Cleaner and coarser-grained, high-angle cross-bedded sandstones have acquired a yellowish to brown weathering coat, while the mud-prone, finer-grained low-angle cross-bedded sandstones have a bluish-grey surface. The weathering color is thus a proxy to distinguish between current-generated sandstones (yellowish) and wave-generated sandstones (bluish).

The first part of this description consists of a tour of faults C1 to C8. The left-hand side compartment of fault C1 consists of finer-grained heterolithics, typical of the lower part of the coarsening-upward cycles; the right-hand side compartment repeats the stratigraphic succession downward, a right-lateral translation of about 3 meters along a nearly straight surface (strike N45° E; dip angle 65° NW). A slice of the left-hand side compartment is in part pulled into the fault plane forming a lens structure, where a high density of deformation bands can be observed (Figure 3.4.4c). Both compartments of the fault display excellent drag features (Figure 3.4.4d).

A second, parallel fault (C2), with a similar sense of displacement is seen some 2.5 m to the right of fault C1; the displacement along the fault plane is about 50 cm (strike N45° E; dip angle 65° NW). A distinct change in the orientation of the fault coincides with the top of the massive sandstone in the middle of the quarry wall, defining an upper and a middle fault segments; a third segment can be observed near the base of the fault, which relays the middle segment past a lens structure.

Further along the outcrop, fault C3 is an antithetic fault, with a sense of displacement opposite to C2 and C1. The left-lateral displacement along fault C3 is in the order of 80-90 cm (strike N55° E; dip angle 65° SE). This fault displays changes in fault plane angle in accordance with the major bed boundaries. A high density of deformation bands can be observed within the sandstone beds, and coaly laminae provide

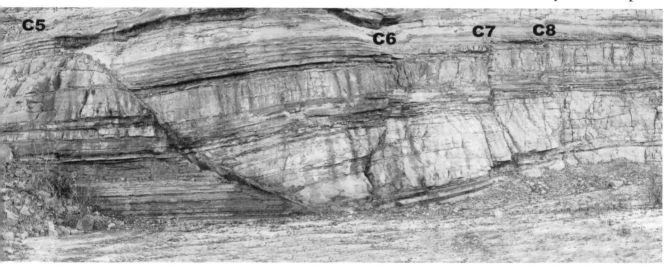

correlation markers across the fault. Faults C2 and C3 merge just below the exposure level (Figure 3.4.4e); from this location the single fault trace can be followed on the quarry floor towards the road. These two intersecting faults with opposite shear sense are called conjugate faults, and between themselves define a small graben structure on the top. At the base of this graben, an 80 cm thick sandstone bed displays sharply dipping laminae, which define the lateral edge of a lobate sand body.

A minor left-lateral normal fault (C4) with a displacement of some 30 centimeters is visible a short distance further (strike N55° E; dip angle 80° NW). On the terrace dominating the quarry wall, faults C4 and C5 can be seen to join up into a single fault trace. Fault C5 (Figure 3.4.4f) is a prominent listric (spoon-shaped), low-angle structure that offsets the right-hand side compartment downwards along a displacement of nearly 5 m (strike N50° E; dip angle 60-45° NW). As a result, the thin-bedded heterolithics of the base of the coarsening upward cycle are in contact with the thicker cross-bedded sandstones of the higher part of the same depositional cycle. The lower part of the fault exposure shows a slice of the left-hand side compartment dragged as a complex rock lens into the fault plane.   The adjacent rim of the sandstone unit is re-cemented into a harder, cm-wide oxide layer. Clay smear is observed all along the fault plane, creating a lateral, impermeable membrane that seals the two adjacent fault compartments.

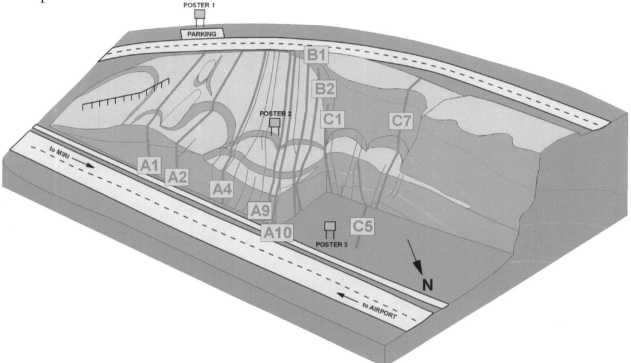

Figure 3.4.4b: Miri Airport Road outcrop's block diagram with fault systems

On the right-hand side of the fault, the downthrown thicker sandstone unit displays textbook examples of herringbone cross stratification, where alternating flow directions create cross-beds with opposite directions (Figure 3.4.4g). These typical tidal depositional environments characterized by bidirectional currents also include examples of escape burrow structures indicating rapid deposition.

The sixth fault (C6) along the main wall of the quarry is a left-lateral normal fault, with a displacement of 30 cm (strike N55° E; dip angle 60° SE). It is followed by fault C7, a subvertical, left-lateral normal fault, extending throughout the quarry, with a displacement of 35-40 cm (strike N55° E; dip angle 65° SE). Faults C6 and C7 extend high up above the higher quarry floor, before they merge together.

Further away, fault C8 is a minor, subvertical, left-lateral normal fault, with a displacement of 30-35 cm (strike N55° E; dip angle 65° SE).

The excursion continues by returning to Poster 3 and observing the quarry wall to the left, where faults B1 and B2 outcrop. These two faults allow observations of deformations in the finer-grained and thinner-bedded unit.

Figure 3.4.4c: Multi-meter offset along fault C1, "456 Sands", Miri Airport Road outcrop

Figure 3.4.4d: Drag features along fault C1, "456 Sands", Miri Airport Road outcrop

Figure 3.4.4e: Antithetic faults C2 and C3, "456 Sands", Miri Airport Road outcrop

Figure 3.4.4f: Antithetic faults C5 and C6, "456 Sands", Miri Airport Road outcrop

Figure 3.4.4g: Tidal sandstones with bi-directional, herringbone cross stratification, "456 Sands", Miri Airport Road outcrop

Fault B1 is the first fault on the left; it is a left-lateral normal fault with a minor offset, in the order of 10 to 20 cm. A high density of deformation bands can be seen in its vicinity. Slightly to the right, fault B2 is a right-lateral normal fault with a vertical offset of about 20-30 cm. These two faults can be seen to merge into one structure at the top of the wall.

The excursion continues in the downhill direction along the walkway, next to the road, where the A-series faults can be observed, from A10 through A1.

Faults A9 and A10 are located close to the road, at the edge of the quarry; fault A10 splits into two steep, diverging segments, defining a larger fault block, which shows little lateral displacement (Figure 3.4.4h). Faults A9 and A10 form together a major and complex, left-lateral normal fault zone, with a vertical offset in the order of 6-7 m. The fault zone includes thick rock lenses lined by clay injection, and prominent deformation bands. The fault zone extends throughout the upper outcrop, where it passes on the side of Poster 2.

Over the next 15-20 meters, four faults (A8 to A5) can be observed, which are characterized by minor vertical displacements (10 to 20 cm each), and well developed lateral damage zone. The deformation bands occasionally show horse-tail structures.

Fault A4 is recognizable as a steep (60°), prominent, right-lateral normal fault, which crosses the whole outcrop wall. It has a vertical offset of 2.5 m. On the terrace above the main road level, fault A4 dissects an older left-lateral normal fault, creating a triangular structure (Figures 3.4.4i,j). A fault zone with larger lens structures can be observed at that location. Halfway up the outcrop, the fault is lined up by a wide damage zone, including separate parallel faults and a network of deformation bands. In the horizontal direction, the fault often displays a striated damage zone, which can reach 1-2m; horse-tail structures are also visible along horizontal cuts.

Some 6 meters further, fault A3 is a steep (about 60°), right-lateral normal fault with a vertical offset of 1.2 m. The sandstones in the fault zone are strongly mineralized, and thus the fault stands out as a sharp break along a narrow zone (typically 5cm).

Some 10 meters further down the roadside path, fault A2 can be seen as a steep (about 60°), right-lateral normal fault, with a vertical offset of about 15 cm. Halfway up the outcrop, the fault splits into a series of parallel faults, spaced some 50 cm apart.

Fault A1 is located some 20 m further away, underneath the highest part

Figure 3.4.4h: Faults A9 and A10 offset an older normal fault visible on the bottom left, "456 Sands", Miri Airport Road outcrop

of the cliff face to the right. It is a steep (about 60°), right-lateral normal fault, with a vertical offset of approximately 40-50 cm. As the fault crosses thicker sandstone intervals, the fault zone widens to about 10 cm, and a striated damage zone reaches 50 cm on each side, with horse-tail structures visible along horizontal cuts.

Figure 3.4.4i: Triangular structure resulting from multiphase extensional-transtensional deforma-tions, "456 Sands", Miri Airport Road outcrop

Figure 3.4.4j: Detailed view of the triangular structure, "456 Sands", Miri Airport Road outcrop

The excursion path crosses successively younger strata on the way downhill (strata dip at angles of 10-12°), exposing a coarsening upward cycle. Heterolithics with hummocky cross stratification are exposed adjacent to the road, characterized by millimeter-thick laminae of coaly debris and clay clasts levels. As the stratigraphically younger sandstones are crossed, the path opens to the right, away from the road. Crossing a short overgrown area, one reaches the base of a handrail, leading to the higher terraces of the outcrop. Polygons of limonitic crusts are seen along the way, and *Ophiomorpha* burrows are exposed alongside the handrail.

The excursion continues in the direction of Poster 2. From the end of the staircase, continue walking on the same elevation. This path will take you towards the edge of the terrace, and will cross horizontally all the faults previously observed in the vertical cut (road wall). Fault A1 is identifiable as the first through-going horizontal track, accompanied by a meter-wide lateral damage zone. The bioturbated, hummocky cross-stratified sandstones of the second coarsening upward cycle are well exposed on the way.

After a stop at Poster 2, continue alongside the same terrace, to reach faults C1 through C8, above the quarry floor where the excursion started (keep a safe distance of two meters or so from the edge of the quarry). These faults can be observed along their strikes here and also in a vertical cut on the left-hand side of the quarry. An interesting exercise consists in trying to identify the individual faults from the observations made at the base of the quarry (displacement direction and magnitude).

The visit continues by returning to Poster 2, and observing the through-going A9 fault at its base (Figures 3.4.4k,l). Note the orientation of the drag features and compare the observations made of the same fault, on the vertical section below.

Walk uphill towards the road following this fault trace on the back of the outcrop and towards Poster 1 (Regional Geology). This location offers an open view towards the South China Sea and towards Lambir Hills.

From your parking area, drive back, taking Jalan Miri Bintulu in the direction of the airport, and enter the Taman Awam parking area for a last overview of the outcrop across Jalan Miri Bintulu (do not attempt to cross this highly transited route). At Taman Awam, one can take the spectacular skywalk and look at the eastern flank of the Miri anticline. The walk provides dazzling views all around and ends at the observation tower, at the entrance of Taman Awam. An additional 30 minutes will be needed to complete this tour.

Figure 3.4.4k: Fault A9 below Poster 2, "456 Sands", Miri Airport Road outcrop

Figure 3.4.4l: Detailed view of fault A9, "456 Sands", Miri Airport Road outcrop

3.4.5 Sealing Faults

Faults commonly occur in petroleum basins and the assessment of sealing or conductive behavior of faults is crucial in understanding the migration and entrapment of petroleum. Fault outcrops in and around Miri provide an opportunity to study the sealing processes of faults; this is especially true for the Miri Airport Road Outcrop where we can view faults in three dimensions. Sorkhabi and Hasegawa (2005) presented a study of fault sealing at this outcrop; the results are summarized here.

A total of eight faults at the Miri Airport Road Outcrop (faults #C1 through C8) are shown in Figure 3.4.5a together with structural data.

An examination of the faults shows that the fault zone is characterized by two structural features (Figure 3.4.5b): (1) shale smear, and (2) deformation bands.

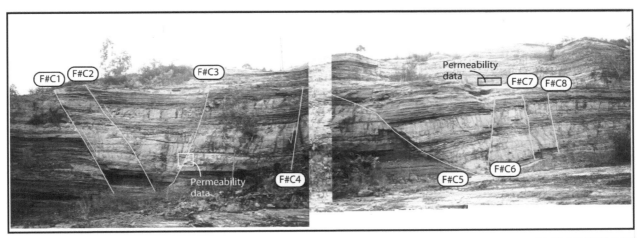

Fault	Strike	Dip Angle	Displacement (cm)	Shale Smear Factor (SSF) (Throw / Shale Layer) (cm)	
#C1	N40-45°E	65°NW	280	250 / 70 = 3.6	Continuous smear
#C2	N40-45°E	65°NW	50	40 / 70 = 0.6 40 / 17 = 2.4	Juxtaposition Continuous Smear
#C3	N55°E	65°SE	85	77 / 80 = 1.0	Juxtaposition
#C4	N55°E	80°NW	30	29 / 80 =1.0	Juxtaposition
#C5	N50°E	60°NW (Steep plane) 45°NW (Listric plane)	500	430 / 80 = 5.4	Continuous smear
#C6	N55°E	60°SE	30	26 / 55 = 0.5	Juxtaposition
#C7	N50-55°E	65°SE	35	31 / 55 = 0.5	Juxtaposition
#C8	N55°E	65-70°SE	35	31 / 55 = 0.6	Juxtaposition

Figure 3.4.5a: View of 8 major normal faults (C1 through C8) together with their structural data (after Sorkhabi and Hasegawa, 2005). Thickest shale layers were selected for measuring the Shale Smear Factor for each fault. Boxes show locations of permeability data presented in Figure 3.4.5f (for fault C6 top view) and Figure 3.4.5e (for fault C3, section view).

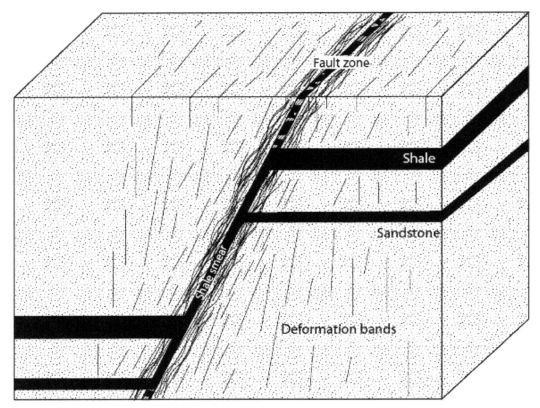

Figure 3.4.5b: A simple block diagram to show the main elements of fault zones in the Miri Airport Road outcrop

Shale (clay) smear along normal faults in sandstone-shale succession has been observed in various sedimentary basins and has been recognized as an efficient across-fault seal. Shale smear in fault zones occurs by several processes as observed in this outcrop (Figure 3.4.5c):

(1) Injection of soft clay into open fractures;
(2) Simple shear involves brittle faulting of mud layers with en-echelon pattern in the fault zone;
(3) Dragging of a parallel set of thin mud layers into fault plane by a combination of shear and injection;
(4) Abrasion whereby the passage of a displaced mud layer leaves a thin veneer of clay on sandstone surface along the fault plane.

The efficiency of shale smear as an across-fault seal depends on its continuity and thickness on the fault plane. This is quantified as Shale Smear Factor (SSF) which is the ratio of fault throw to the thickness of source shale layer. SSF values smaller than 1 imply juxtaposition of sandstone against the displaced shale layer. Several studies of outcrops and oil fields from around the world indicate that SSF values up to 6 give continuous shale smears which may seal the fault for across-fault oil migration because clay material is impermeable compared to sandstone. In all of the faults observed in the Miri Airport outcrop, thin layers of black shale are present on the fault planes indicating the continuity of shale smear.

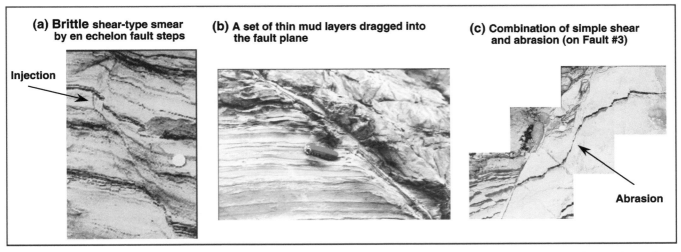

Figure 3.4.5c: Various styles of shale smearing observed in the Miri Airport Road outcrop (after Sorkhabi and Hasegawa, 2005)

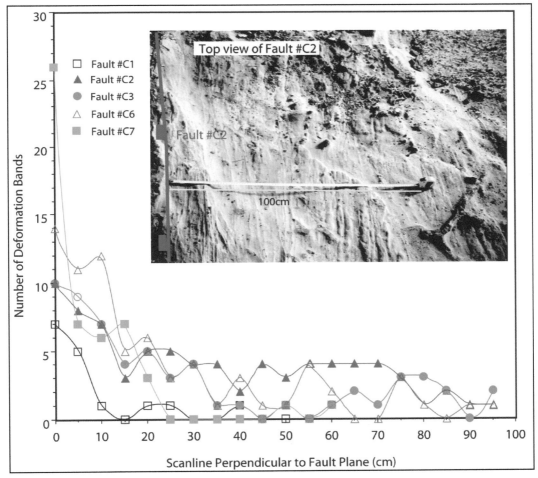

Figure 3.4.5d: Increase in the number of deformation bands in the Miri Formation sandstone as one approaches the fault (data were taken along scan lines perpendicular to fault plane C1 through C7). The inset photo show an example for fault C2 (after Sorkhabi and Hasegawa, 2005)

Deformation bands are thin (mm wide), planar features in faulted sandstone rocks first documented in Utah in the 1970s., and have since been observed in fault outcrops in other parts of the world. Deformation bands are associated with normal faulting in porous sandstone and are the smallest visible shear fractures where the host sandstone has undergone fracturing, crushing, and porosity collapse.

In the Miri Airport Road Outcrop, one can observe that the frequency of deformation bands increases toward the fault plane (Figure 3.4.5.d). In the vicinity of

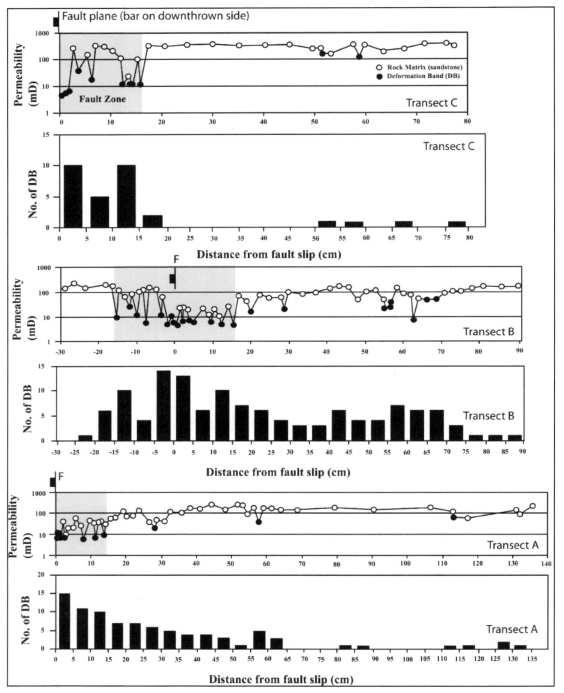

Figure 3.4.5e: Measurements of intensity of deformation bands and changes in permeability (millidarcy) along three transects parallel to fault C6 (top view, location shown in Figure 3.4.5a). Permeability measurements were taken on the outcrop using a probe minipermeameter and nitrogen gas (after Sorkhabi and Hasegawa, 2005)

the fault plane, the deformation bands are so closely spaced that they form a zone of deformed rock (fault zone). In the study area, this width of the fault zone is about 10 cm on one side of the fault.

Deformation bands are known to decrease the permeability across the rock. This phenomenon is demonstrated for two fault zones of the Miri Airport Road outcrop: Fault #C6 shows (Figure 3.4.5e) where the frequency of deformation bands and permeability of both deformation bands and sandstone rock were measured for three transects, and Fault #C3 (Figure 3.4.5f) where the measurements were made along four transects. Permeability data (measured in millidarcy units) depicted in Figures 3.4.5e and 3.4.5f indicate that (1) the permeability of individual deformation band is lower than that of the adjacent sandstone; (2) the permeability of fault zone characterized by a closely-spaced zone of deformation bands is much lower than that of the host sandstone away from the fault zone. Overall, deformation bands in the fault zone have 10 times lower permeability than the undeformed sandstone reservoir rock away from the fault zone. Note that the permeability measurements were taken at the outcrop by a probe minipermeameter and using nitrogen gas.

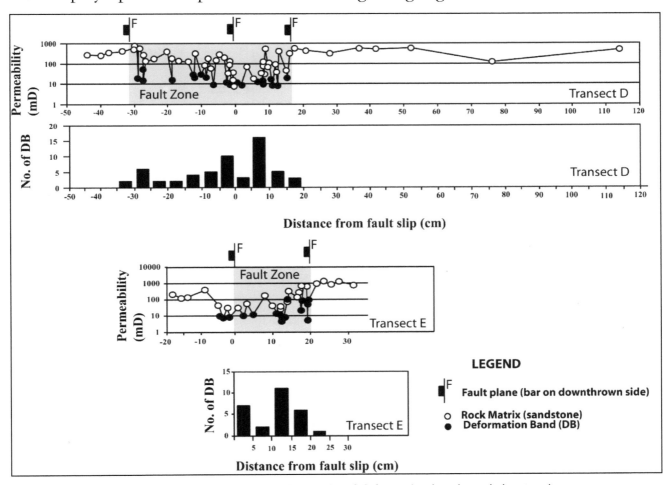

Figure 3.4.5f: Measurements of intensity of deformation bands and changes in permeability (millidarcy) along two transects parallel to fault C3 (section view, location shown in Figure 3.4.5a). Permeability measurements were taken on the outcrop using a probe minipermeameter and nitrogen gas (after Sorkhabi and Hasegawa, 2005)

3.5 Canada Hill Thrust, Hospital Road

3.5.1 Access and Location

From the Clock Tower roundabout at km 0.0, follow Jalan Miri Bintulu southward (towards the airport). The road climbs a small hill, and after 1.3 km a bifurcation to your left will take you to Jalan Cahaya, the Hospital Road. Drive further 2.2 km; pull on the left-hand side of the road and park your vehicle underneath one of the trees lining up the road. The parking spot should be at the edge of the Miri Hill on the left, before the flat area that leads towards the General Hospital (N 04° 22.649'; E 113° 59.451', Figure 3.5.1a).

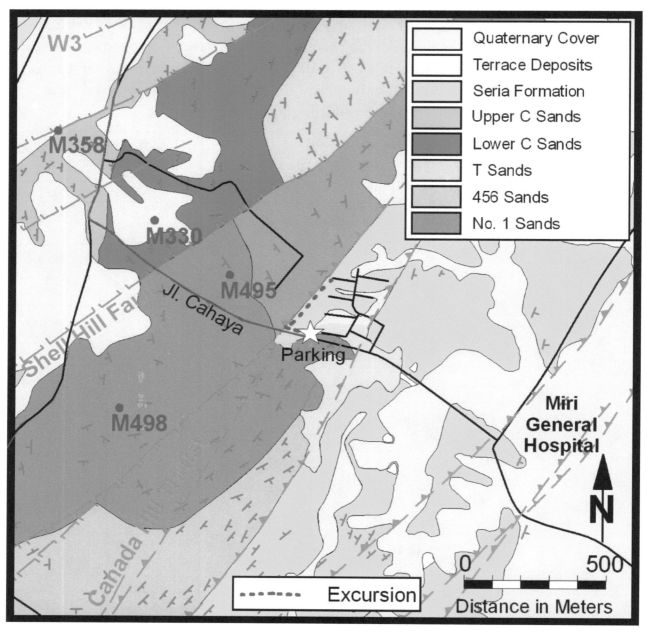

Figure 3.5.1a: Location and geological map of the Canada Hill Thrust outcrop, Hospital Road (M330, M358, M495, M498: location of Miri wells)

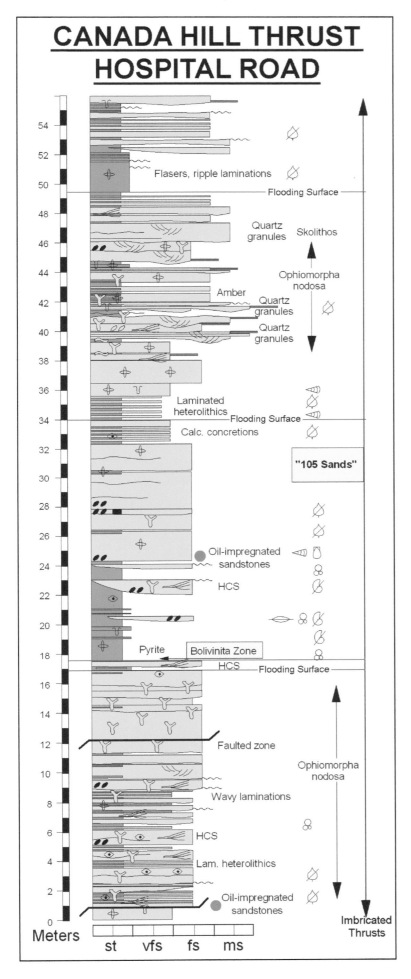

CANADA HILL THRUST
HOSPITAL ROAD

Flasers, ripple laminations

Flooding Surface

Quartz granules — Skolithos

Ophiomorpha nodosa

Amber — Quartz granules

Quartz granules

Laminated heterolithics

Flooding Surface

Calc. concretions

"105 Sands"

Oil-impregnated sandstones

HCS

Pyrite — Bolivinita Zone

HCS — Flooding Surface

Faulted zone

Ophiomorpha nodosa

Wavy laminations

HCS

Lam. heterolithics

Oil-impregnated sandstones

Imbricated Thrusts

Meters

st vfs fs ms

Figure 3.5.3a: Stratigraphic log of the "105 Sands", Canada Hill Thrust on Hospital Road

89

3.5.2 Logistics and Safety

This is a one hour-long stop along a prominent cliff of the Miri anticline. Accessing the outcrop from the parking area is sometimes a little difficult, due to the crossing of a ditch, covered with high grass. Parts of the cliff are very steep and should not be approached. As the area is under development, conditions for a geological tour will vary with time.

3.5.3 Outcrop Highlights

- The Canada Hill Thrust
- The "105 Sands", the second largest reservoir contributor to the Miri field production (Figure 3.5.3a).
- Oil seeps along the fault
- Vertical beds of the Seria Formation

3.5.4 Geological Description

From your parking location, walk towards the base of the cliff and here you will encounter the Canada Hill Thrust (Figure 3.5.4a), displaying a 4-meter wide fault zone, which contains a number of rotated lenses. The sandstone lenses are more competent and rigid, forming large meter-size units, while the heterolithic lenses are smaller, decimeter-size units. All these lenses are embedded in darker mudstone /claystone matrix, which is the sealing feature of the fault. It is likely that much of the claystone is smeared from the overlying shale units or some of it may even be injected into the fault zone from the underlying Setap shales.

Figure 3.5.4a: Complex damage zone along the Canada Hill Thrust plane, Hospital Road outcrop

In following the thrust contact along the base of the cliff, you will soon reach a small waterfall (Figures 3.5.4b). Here you will encounter a complex reverse-fault zone; the waterfall follows this zone of deformation, characterized first by kink bands and tight chevron-style folds (Figure 3.5.4c), and further away by meter-scale box-folds and normal folds (Figure 3.5.4d). There is usually a pervasive smell of petroleum in the vicinity of the Canada Hill Thrust and oil-impregnated sandstones can be seen within and above the fault zone.

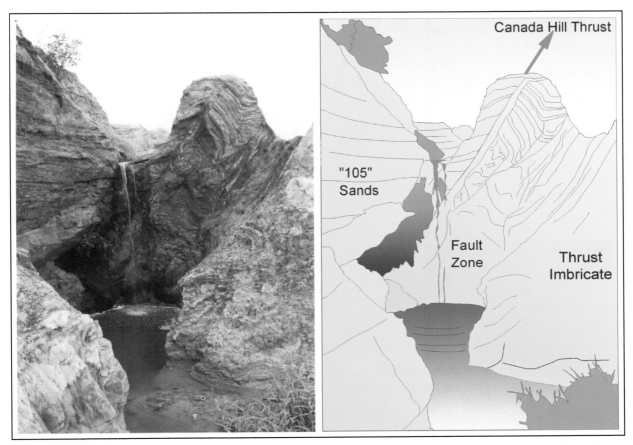

Figure 3.5.4b: Small water stream and cascade along the Canada Hill Thrust plane, Hospital Road outcrop: Photograph and interpretative diagram

The Canada Hill Thrust is a major fault, striking N40° E, along the southeastern edge of the Miri anticline, but its best exposure is at this outcrop (Figure 3.5.4e). Cross the little stream with care, pass along a sandstone ledge and turn left to reach a flat area facing west towards the cliff. At this location, one is standing on vertical strata striking about 30° and looking at the gently tilted beds on the other side of the fault zone. Based on micropaleontological analyses, the underlying vertical strata correlate with the Seria Formation.

Organic-rich, centimeter-bedded strata of the Seria Formation can be observed (Figure 3.5.4f), which are offset by a large number of small extensional faults, striking approximately 330°, perpendicular to the orientation of the Canada Hill Thrust (Figure 3.5.1a). There are few exposures of the footwall block (as it is highly weathered or buried under the construction), consisting of the Seria Formation but the cliff of the upthrown (hangingwall) block provides a spectacular cut through the "105 Sands" of the Miri Formation.

Figure 3.5.4c: Monolith of intensely folded strata within
the Canada Hill Thrust zone, Hospital Road outcrop

Figure 3.5.4d: Box fold (now bulldozed away) along the Canada Hill Thrust plane,
Hospital Road outcrop

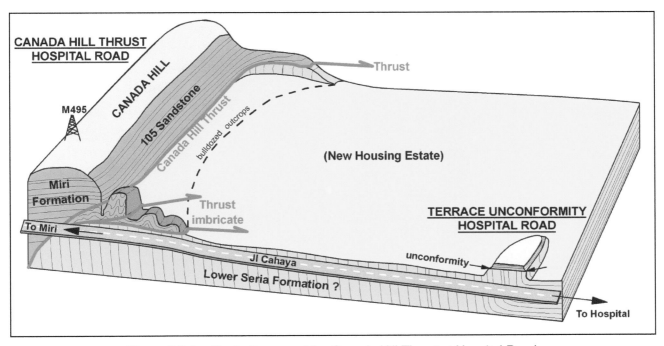

Figure 3.5.4e: Block diagram of the Canada Hill Thrust at Hospital Road

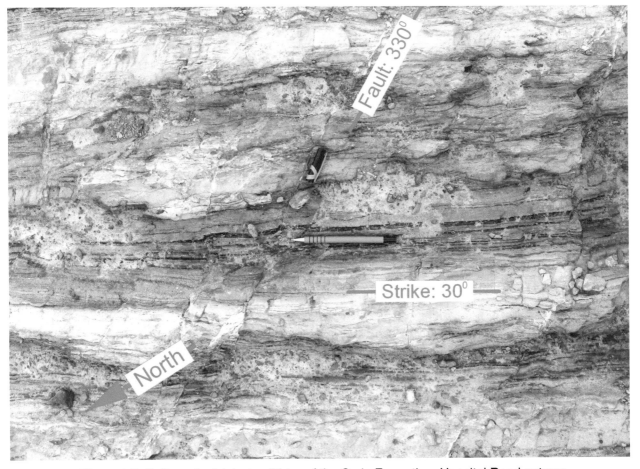

Figure 3.5.4f: Organic-rich heterolithics of the Seria Formation, Hospital Road outcrop

Continue your walk following the fault zone in a NE-direction for about 200 meters. This direction is opposite to the plunging nose of the anticlinal structure, and younger rocks become visible at the base of the cliff. The thrust plane follows the very base of the hill and is not exposed along this portion of the outcrop.

The stratigraphy of the upthrusted block is fairly continuous alongside the cliff (Figure 3.5.4g):

1. At the base, thick interbeds of mudstones and sandstones (5-6 m), pass upward into
2. Bedded sandstones with thin mudstone layers (4 m), passing upward into
3. Interbedded unit of mudstones and sandstones (3 m), showing a vertical increase in the presence and thickness of sandstones beds
4. Thicker sandstone unit (2.5 m), draped by a 30 cm-thick bed of mudstone which is a good marker bed throughout the outcrop
5. Slightly muddy sandstones passing upward into cleaner sandstone beds (6 m thick)
6. Mudstone unit, 7-8 m thick, with a fining upward trend of thin sandstones at the base, and a coarsening upward trend of isolated sandstone beds at the top
7. Sandstones with an 8 m thick fining upward trend, sharp-based and making up the top of the hill.

The section can be followed up for another 25 m, as shown on the stratigraphic column in Figure 3.5.3a.

Unit 7 outcropping near the parking area is possibly part of the "No.1 Sand".

Figure 3.5.4g: Stratigraphic units of the "105 Sandstones", Hospital Road outcrop

Unit 6 is best studied from a distance, as it is difficult to access. The mudstones are marine, offshore deposits, as indicated by its fossil content; the recovered benthonic and planktonic foraminifera correlate with the local "*Bolivinita* Biozone", a marker zone characterizing the top of the "105 Sands". The base of the mudstones makes a sharp boundary with the underlying thick sandstones, a typical flooding surface above which the depositional environment becomes abruptly deeper and more distal. The lower half of the mudstone unit encases occasional decimeter-thick sandstone beds, mainly at its base, followed by thin, centimeter-thick sandstones, representing the more distal depositional environment. The upper half of the mudstone unit is characterized by a vertically increasing number of decimeter-thick sandstone beds; the beds thicken and thin laterally and show characteristic gutters. Hummocky cross stratifications can be recognized within these sandstones; their sharp base with clay clasts and fossil concentration indicate a storm-related origin.

The contact between Units 6 and 7 is sharp and represents a return to a much shallower marine environment.

Continue walking towards the cliff, to make detailed observations of the sandstones, part of the lower stratigraphic succession. Watch for a small gully running parallel to the base of the outcrop.

The amalgamated sandstones of Unit 5 are characterized by cross-bedded, laminated intervals, with aligned clay clasts at the base, and by massive, bioturbated intervals, with abundant *Ophiomorpha* burrows (Figure 3.5.4h). These sandstones were deposited in a shallow marine to coastal, high-energy environment.

Further along the section, one reaches the 30-centimeter thick marker bed separating Units 5 and 4. It consists of finely laminated heterolithics, indicating deposition under low energy conditions, probably in a tidal environment. Note the sharp base of this unit.

The underlying sandstone Unit 4 consists of a number of stacked channel units, downcutting into each other, as can be observed through the entire length of the outcrop (Figure 3.5.4i). Thin laminae of organic-rich mudstones are pervasive throughout the sandstones, a possible indicator for a tidal environment. The base of the unit is an erosional surface truncating in places the underlying unit of sandstone and mudstone alternations (top of Unit 3). Unit 2 consists of massive, bioturbated sandstones, with abundant *Ophiomorpha* burrows. Unit 1 consists of interbedded sandstones and mudstones; its base and the contact with the Canada Hill Thrust are covered by rock debris.

The excursion continues in the direction of a black-and-white painted handrail encircling a water canalization ditch, next to a new housing complex (Figure 3.5.4j). The damage zone of the Canada Hill Thrust is visible within the ditch: it consists in a 4-5 m wide zone with a number of lenses rotated at varying angles, occasionally separated by decimeter-thick mudstones and claystones.

From this location, the Seria Formation outcrops again, dipping at an angle of 50-55° and striking at 30-35° E (Figure 3.5.4k). Alternations of decimeter-bedded bioturbated sandstones and mudstone form part of this footwall block.

The excursion ends at this location; follow the same path to return to the car. The Terrace Unconformity further along Hospital Road can be visited next.

Figure 3.5.4h: Burrowed sandstone bed and *Ophiomorpha* trace fossils, "105 Sandstones", Hospital Road outcrop

Figure 3.5.4i: Stacked, down-cutting channel sandstones, "105 Sandstones", Hospital Road outcrop

Figure 3.5.4j: Sub-horizontal Miri Formation sandstones overthrusting sub-vertical series of the Seria Formation, Hospital Road outcrop

Figure 3.5.4k: Canada Hill Thrust and underlying, sub-vertical sediments of the Seria Formation, Hospital Road outcrop: Photograph and interpretative diagram

3.6. Terrace Unconformity, Hospital Road

3.6.1 Access and Location

From the Clock Tower roundabout at km 0.0, follow Jalan Miri Bintulu southward (towards the airport). The road climbs a small hill, and after 1.3 km a bifurcation to your left will take you to Jalan Cahaya, the Hospital Road. After 2.7 km total distance, pull on the left-hand side of the road and park your vehicle on the opposite side of the bus stop (N 04° 22.636'; E 113° 59.528', Figure 3.6.1a).

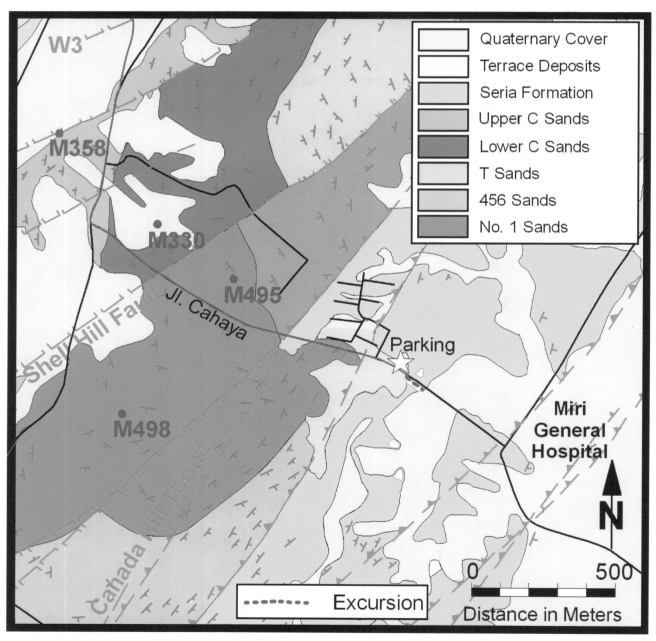

Figure 3.6.1a: Location and geological map of the Terrace Unconformity outcrop, Hospital Road

3.6.2 Logistics and Safety

This is a short stop along a small, but very interesting outcrop. There is no dedicated parking space at the location, so care has to be exercised to find a safe location to park along the road and not impede public transportation.

The location enjoys a good visibility on the through-going traffic.

3.6.3 Outcrop Highlights

• Vertical beds of the Seria Formation
• A spectacular angular unconformity at the base of terrace deposits with white quartzitic sands at the top

3.6.4 Geological Description

Cross the street with caution and reach the outcrop behind the bus stop. Here, the Seria Formation is an alternation of vertically standing mudstone and sandstone beds that can be followed along the small cliff over a distance of some 30 meters. This formation is truncated by the Terrace Unconformity and capped by very fine-grained sandstones, mostly unconsolidated, including a pervasive white sand level at the top (Figure 3.6.4a). The Seria Formation can be further seen outcropping to the right of the road, in the direction of the General Hospital. Before World War II, this formation was sampled through a large number of shallow boreholes, and foraminifera typical for the Seria Formation were described (Artis, 1941).

The outcrop is located between the "Canada Hill Thrust" and the "Inner Kawang Thrust", in an area characterized by near-vertical dips, striking 215-220°. The identification of the top and the bottom of beds is a common stratigraphic problem in very steeply dipping strata. The presence of channelized sandstones, with downcutting lower boundaries is one of the criteria that can be used in this outcrop to determine the original stratigraphic architecture. Figure 3.6.4b shows the erosive base of a sandstone bed, indicating that the sequence is younging towards the General Hospital. The presence of clay clasts at the base of sandstone units is another indicator of the depositional orientation.

The angular unconformity at the base of the brown and overlying white sands terrace deposits (informally, the "lower" and "upper" terrace sandstones) indicates a larger time gap, during which the Late Miocene marine mudstones and sandstones of the Seria Formation were folded and eroded, and ultimately truncated by a marine terrace following the 120 feet-height contours. The overlying very fine-grained white sands ("upper" terrace sandstones) are severely leached, resulting in near clean quartzitic deposits, devoid of any fossils. These white, clean and well-sorted very fine-grained sandstones are present all along the coast of Miri towards Bekenu. Along the coast of Brunei and particularly near Tutong, these deposits are of glass-sand quality (Wilford, 1961), in places reaching near-pure quartzitic composition (99.78% SiO_2).

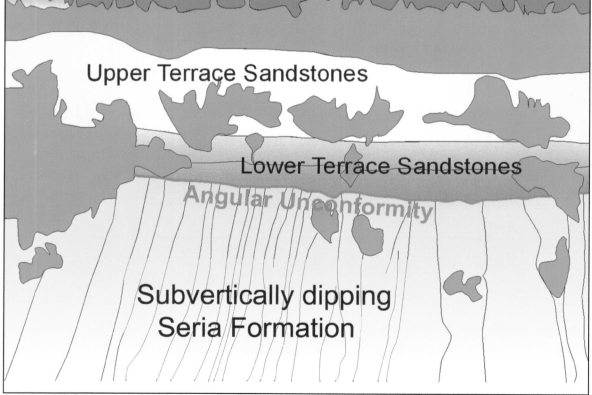

Figure 3.6.4a: Sub-vertically dipping, lithified sandstones of the Seria Formation
truncated and unconformably overlain by unconsolidated, leached sandstones:
Photograph and interpretative diagram. Terrace Unconformity outcrop, Hospital

There is no definitive age determination for these sediments, but a Pleistocene age is most likely (Wilford, 1961). The soils associated with these white sands are extremely poor in nutrients, resulting in a unique floral association, the Kerangas forest, which characterizes a larger part of the northwest coasts of Borneo. Pitcher plants are frequently observed in this environment. Lack of sedimentary structures in the white sands leaves the question of its origin wide open. The underlying brown sands are generally characterized by high angle cross beds, indication a sub-aqueous deposition, in a high-energy environment. These are probably remnants of marine deposits responsible for the down cutting of the terraces in the folded Miocene sediments. The pervasive brown color of these lower sands is due to the filtration of peat-rich deposits. The overlying white sands are possibly marine sandstones deposited in a similar depositional environment,

Figure 3.6.4b: Channel sandstones in the Seria Formation; Terrace Unconformity outcrop, Hospital Road

and subsequently leached. Another, yet untested hypothesis is that of a wind-blown origin.

This outcrop visit can be extended on the same roadside, in the direction of the General Hospital. A newly cleaned area, in the level foreground shows a number of channelized sandstones, separated by mudstone intervals, in an overall tidal-influenced depositional environment (Figure 3.6.4c).

Figure 3.6.4c: Massive, amalgamated channel sandstones in the Seria Formation (yellowish and reddish color), with discontinuous beds of silty shales (bluish color). Note the white sandstones of the terrace deposits in the background. Terrace Unconformity outcrop, Hospital Road

3.7 Jalan Lopeng Quarry

3.7.1 Access and Location

From the Clock Tower roundabout at km 0.0, follow Jalan Miri Bintulu southward (towards the airport). The road climbs a small hill, and after 1.3 km a bifurcation to your left will take you to Jalan Cahaya, the Hospital Road. After 3.1 km total distance, at the roundabout in front of the hospital, turn left and follow Jalan Padang Kerbau northward along the Miri Hill. After 5.5 km total distance, turn left into Jalan Oil Well #1 and turn left again after some 50 meters. Pass the old quarry and park on the right-hand side of the road, next to the outcrops at the base of the quarry (N 04° 23.489'; E 113° 59.995', Figure 3.7.1a).

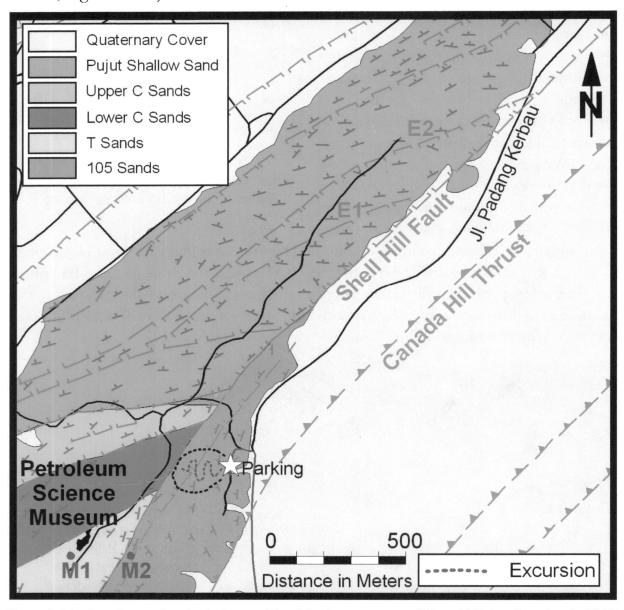

Figure 3.7.1a: Location and geological map of the Jalan Lopeng quarry (M1 and M2: location of wells Miri#1 and Miri#2)

3.7.2 Logistics and Safety

This is a one- to two-hours stop, mainly to climb to the top of the quarry. There is no marked path, and the approaching area next to a pond is sometimes overgrown: be careful to remain on high ground all the time. Part of the old quarry has recently been used as a storage area and access conditions may change at short notice.

The visit of the quarry does not present special dangers, with the exception of the top-most part, a vertical cliff at the base of a mudstone unit. Climbing in this part of the cliff requires special precautions, and should only be undertaken by experienced, careful geologists, in good weather conditions. Put on a helmet for this part of the visit.

3.7.3 Outcrop Highlights

• Exposure of the "105 Sands", the second largest reservoir contributor to the Miri field production (Figure 3.7.3a)
• Detailed sedimentological observations of shallow coastal marine and tidal depositional environments
• Presence of a shallow marine fossil fauna

3.7.4 Geological Description

From the parking area, the main quarry wall is seen to extend high up toward the top of the Miri Hill (Figure 3.7.4a). To start the visit, walk up towards the quarry, passing on the right side of a small pond. To the right, the lower part of the quarry is not anymore accessible, as it is now used as an industrial storage area. The lowermost sandstone unit exposed in the quarry is

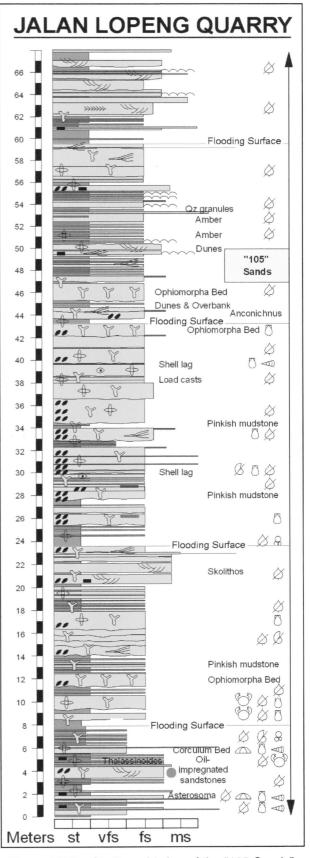

Figure 3.7.3a: Stratigraphic log of the "105 Sands", Jalan Lopeng quarry

103

characterized by clay clasts and lithoclasts (Figure 3.7.4b) and abundant *Dactyloidites* trace fossils (Figure 3.7.4c).   A prominent sandstone ledge about 1m-thick can be traced at the base of the quarry (Figure 3.7.4d); stacked tidal bars can be recognized in the upper part of this ledge (Figure 3.7.4e). A large part of these sandstones is impregnated with petroleum, and a distinct smell of crude oil is pervasive throughout the lower part of the quarry.

Figure 3.7.4a: Overview of the Jalan Lopeng quarry

Figure 3.7.4b: Bioturbated basal sandstone bed with clay clasts, "105 Sands", Jalan Lopeng quarry

Figure 3.7.4c: Radial burrow structures *(Dactyloidites)* in basal sandstone bed, "105 Sands", Jalan Lopeng quarry

Continue walking up the edge of the quarry, crossing an 8-9 meter thick unit of silty mudstones and heterolithics, with bioturbated sandstones including laminae of carbonaceous material. Some of the sandstone beds are stained with petroleum.

Climb onto the first terrace, following a small path at the edge of the outcrop, on the left. Follow the terrace all the way, crossing to the right side of the quarry. At this location, intensively bioturbated grey sandstones capped with reddish-colored, mineralized sandstones outcrop and a pervasive smell of hydrocarbon will be noticed. Various types of burrows can be observed here, including large vertical cylindrical traces (Figure 3.7.4f), smaller, smooth meandering traces back-filled with pellets, and horizontally branching, low-curvature *Thalassinoides* (Figure 3.7.4g), occasionally with a hard, mineralized "shadow zone" at their base (Figure 4.1.4c); fossil shells, occasional bivalves, sea urchins and crabs (Figure 3.7.4h) are also found.

The excursion continues up the gentle slope of the quarry, climbing a series of meter-thick cycles characterized by quiet-water, bioturbated heterolithics capped by high-energy sandstones. The sandstones are sometimes amalgamated, as can be recognized by dividing layers of aligned clay clasts (Figure 3.7.4i); the lateral thickening and thinning and the slightly coarser grain size (up to medium grained sandstones) indicate that the sandstones correspond to a system of elongated channels. Going up-section, the presence of thin clay drapes over the sandstones is an indication that deposition was occurring under tidal conditions.

Figure 3.7.4d: Meter-thick sandstone ledge at the base of the quarry, "105 Sands", Jalan Lopeng quarry

Look for a thick, yellowish, coarser sandstone bed with distinct high angle cross stratification and follow its base, backwards to the left for some 20 meters, until a small cliff is reached, exposing the whole unit(Figure 3.7.4j).

A thinning-upward sequence of heterolithics separated by centimeter-thick, fine-grained sandstone beds is observed at the base of the thick sandstone unit. Ripple marks characterize the top of the sandstone beds indicating gentle currents in an overall low-energy, shallow-water, tidal depositional environment (Figure 3.7.4k). The overlying sandstones are coarser-grained, have bedforms and high-angle cross beds with clay clasts at the base, and are occasionally draped by mudstone laminae; they indicate an abrupt change to high-energy conditions within a tidal environment. The contact is largely non-erosive, which can imply an increase in accommodation space, possibly linked to concomitant extensional faulting on the inner shelf.

The 2 m thick sandstone unit is generally medium-grained and is an amalgamation of a series of sandstone bars, often channelized, with occasionally preserved thin mudstone and coaly drapes characterizing an abandonment facies. *Ophiomorpha* and other smooth vertical burrows are present within the unit, which may be part of a distributary channel within a tidal environment.

Continue walking up on the right side of this small cliff. A low-angle laminated sandstone with *Ophiomorpha* burrows capped by a limonitic surface is seen overlying the coarser sandstones described above. They are, in turn, overlain by silty mudstones and heterolithics, creating a slight recess in the slope profile.

Continue upwards on a path that follows the edge of a zone with shrubs. This section consists of an alternation of heterolithic and bioturbated sandstone units, soon dissected by a series of shear faults striking at N 40-50° E (Figure 3.7.4l). These faults are running parallel to the Canada Hill Thrust, which marks the eastern edge of Miri

Miri Hill along this segment.

The magnitude of the throw along the reverse Canada Hill Thrust and Shell Hill Fault can be assessed when one considers that the sandstones outcropping in the quarry are buried some 350-400 meters below sea level only a kilometer away, vertically below Miri Well #1.

Passing the fault zone, the section consists again of an alternation of bioturbated heterolithics and fine-grained sandstones with coaly laminae and coal clasts (Figure 3.7.4m). Fossils of marine bivalves and gastropods are found at the base of some of the sandstone beds.

Figure 3.7.4e: Stacked tidal bars, with erosional contacts, "105 Sands", Jalan Lopeng quarry

Figure 3.7.4f: Vertical tube-like burrow *(Skolithos)* in high-energy sandstone bed, "105 Sands", Jalan Lopeng quarry

Figure 3.7.4g: Branching burrows with smooth walls *(Thalassinoides)*, "105 Sands", Jalan Lopeng quarry

Figure 3.7.4h: Crab embedded in bioturbated silty sandstone, "105 Sands", Jalan Lopeng quarry

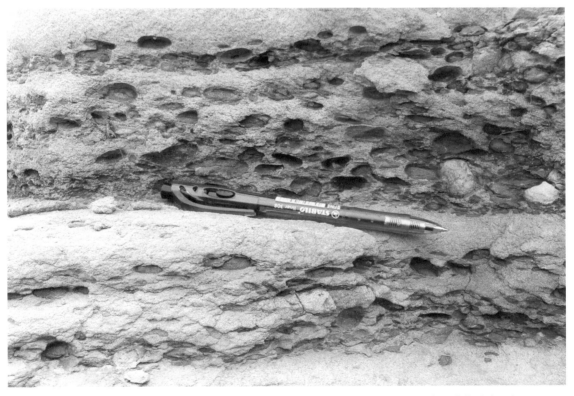

Figure 3.7.4i: Stacked sandstone beds with imbricated clay clasts, "105 Sands", Jalan Lopeng quarry

Figure 3.7.4j: Ledge of medium- to coarse-grained sandstones with yellowish color, "105 Sands", Jalan Lopeng quarry

Figure 3.7.4k: Horizon with ripple marks below pen, "105 Sands", Jalan Lopeng quarry

Figure 3.7.4l: Shear fault at the upper part of the "105 Sands", Jalan Lopeng quarry (Note the right-lateral sense of displacement)

The excursion reaches a high terrace in the quarry, where abundant trace fossils can be observed on bedding planes. The back of the terrace consists in a 4-5 m rock face, with 2 cycles of heterolithics and sandstones. The lower cycle shows a vertical increase in the frequency of sandstone beds, many of which have well preserved symmetrical rippled tops with a coat of limonite. The interbeds consist of thin mudstone drapes, and indicate a low-energy, tidal depositional environment. The higher part of the sandstone unit is a series of high-angle cross beds downcutting into each other, and small sand bars, all indicative of a higher energy depositional environment.

A 10 cm thick, medium-grained sandstone overlies the higher sedimentary cycle and marks the base of a short terrace. The section continues with alternating heterolithics and sandstone beds. A 1-2 cm thick layer of coal that may indicate partial emergence abruptly overlies 20 cm thick bioturbated sandstones of shallow marine origin. Higher up, a 1 m thick, bioturbated sandstone unit is overlain by a mudstone unit, at the foot of the top-most rock face of the quarry. Note the safety recommendations regarding this section of the outcrop.

The higher cliff shows an alternation of laminated and decimeter-bedded heterolithics and downcutting sandstones (Figure 3.7.4n); these are capped by mudstone drapes and ripple marks, indicating continued deposition in a tidal environment under fluctuating energy conditions. A prominent medium-grained sandstone with high-angle cross beds is visible in the lower third of the cliff face. The transition from the sandy foresets to the laminated bottomsets can be observed alongside this sandstone bed.

The outcrop visit finishes at this point; return to your vehicle following the same path.

Note that well Miri#2 was drilled within this formation (Figure 3.7.1a) and was found to be dry! (see Chapter 1.4.2).

Figure 3.7.4m: Bioturbated heterolithics with coal clasts, "105 Sands", Jalan Lopeng quarry

Figure 3.7.4n: Tidal succession characterized by sand-shale layering, ripple marks, laminated heterolithics, coarse foresetted sandstones (tidal bars) and shallow channel incisions, "105 Sands", Jalan Lopeng quarry

3.8 Jalan Padang Kerbau Outcrop 1, near Lot 503

3.8.1 Access and Location

From the Clock Tower roundabout at km 0.0, follow Jalan Miri Bintulu southward (towards the airport). The road climbs a small hill, and after 1.3 km a bifurcation to your left will lead you to Jalan Cahaya, the Hospital Road. After 3.1 km total distance, at the roundabout in front of the hospital, turn to the left and follow Jalan Padang Kerbau back northward along the Miri Hill. Pass Jalan Oil Well #1 for another 600 m and after 6.1 km total distance, park on the left-hand side of the road, next to a short road leading left to Lot 503 (Figure 3.8.1a).

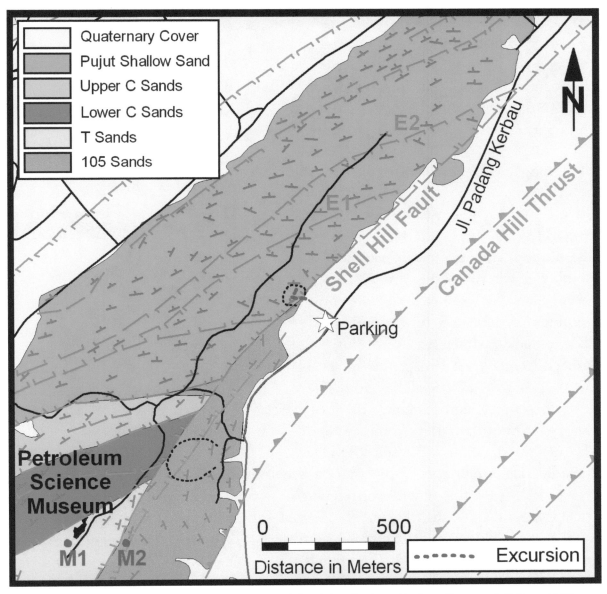

Figure 3.8.1a: Location and geological map of the Jalan Padang Kerbau Outcrop 1 (M1 and M2: location of wells Miri#1 and Miri#2)

3.8.2 Logistics and Safety

This is a half-hour stop to visit a small quarry face, situated at the back of a private house. Ask permission from the owner to walk on the property. The visit of the quarry does not present special dangers, with the exception of the side areas, which are very steep and should not be approached. The geological features of interest are in the central part of the outcrop and are safe to approach.

3.8.3 Outcrop Highlights

• Exposures of the Pujut Shallow Sands (Figure 3.8.3a)
• Intra-formational angular unconformity

3.8.4 Geological Description

From the parking area, walk up the road leading to Lot 503, and with the permission of the home owner, go to the backside of the building, where an old quarry face is exposed. There, a sand-rich section some 40 m thick is dissected by a number of smaller extensional faults; according to Schumacher's geological map, this interval forms part of the Pujut Shallow Sand. The base of the quarry exposes some 10 m of stacked, bioturbated, fine-grained sandstones, with internal layers of clay clasts indicating a high-energy depositional environment (Figure 3.8.4a).

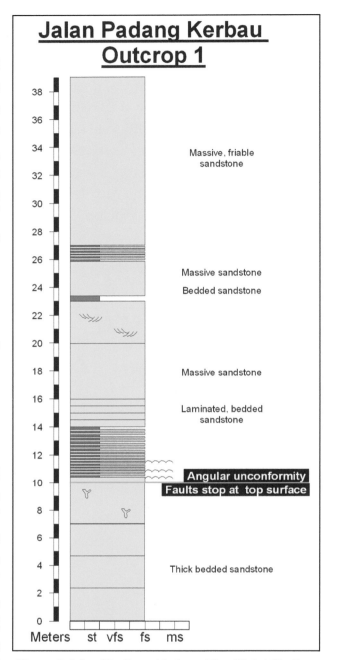

Figure 3.8.3a: Stratigraphic log of the "Pujut Shallow Sands", Jalan Padang Kerbau Outcrop 1

The top of this unit is truncated at a low angle unconformity and overlain by an onlapping unit of slightly coarser-grained sandstone beds with a characteristic yellowish color. This unconformity is best observed from the centre right part of the quarry, looking towards the left (Figure 3.8.4b); it can also be observed along a steep cliff further north (Figure 3.8.4c). The angular contact between the two units is a proof that the lower unit was slightly rotated and eroded, prior to the deposition of the upper unit. Some extensional faults abruptly end at the unconformity (Figure 3.8.4c), while all faults terminate at the base of the overlying sand-shale interbeds (Figure 3.8.4d). There is no evidence for emergence of the underlying sandstones prior to being capped by the coarser, onlapping sandstones above the unconformity. Such block rotations simultaneous with the deposition of the Miri Formation are likely due to lateral contractions and withdrawal of the underlying water-bearing Setap shales, reacting to the increased pressure of the newly accumulated sediments (Figure 3.8.4e).

Figure 3.8.4a: Overview from the base of the quarry, "Pujut Shallow Sands", Jalan Padang Kerbau Outcrop 1

Above the unconformity and healing the angular contact, the transgressive unit consists in cross-bedded, fine to medium-grained sandstones of thicknesses varying from 30 to over 50 cm. Where the sandstone is thickest, a number of rippled surfaces can be seen within the unit; these surfaces vanish laterally as the thickness of the sandstone bed decreases. The upper surfaces of all sandstone beds are rippled.

Rubble

Figure 3.8.4b: Angular unconformity within the "Pujut Shallow Sands", Jalan
Padang Kerbau Outcrop 1: Picture and interpretative diagram

Overlying this healing unit, a 30 cm thick unit of muddy heterolithics creates a small recess and is a good marker bed throughout the quarry (Figure 3.8.4a); it indicates an episode of deposition under lower energy conditions. This softer unit is overlain by a 10 m thick unit of laminated, bedded sandstones, followed by massive, bioturbated sandstones and topped by high-angle cross-bedded sandstones. The stratigraphic architecture of the higher parts of the quarry shows a repeat of this cycle. The visit to this outcrop ends at this point; it is time to return to your vehicle.

Figure 3.8.4c: Angular contact between massive sandstones at the base and tidal sand-shale interbeds at the top, as seen on a higher cliff face, "Pujut Shallow Sands", Jalan Padang Kerbau Outcrop 1

Figure 3.8.4d: Fault terminations in sandstones below the angular unconformity or within the overlying sheared sandstone beds, "Pujut Shallow Sands", Jalan Padang Kerbau Outcrop 1

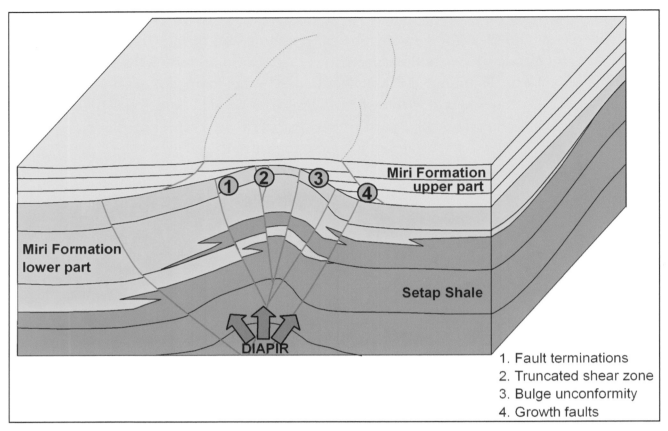

Figure 3.8.4e: Interpretative block diagram of syn-sedimentary deformation structures within the Miri Formation

Note that Schumacher's geological map (Figure 3.8.1a) indicates that the Shell Hill Fault apparently crosses the base of this outcrop, where it offsets the "Pujut Shallow Sand" against the "105 Sands". However, there is no indication in the outcrop for the presence of the Shell Hill Fault at this location.

3.9 Jalan Padang Kerbau Outcrop 2, near Lot 447

3.9.1 Access and Location

From the Clock Tower roundabout at km 0.0, follow Jalan Miri Bintulu southward (towards the airport). The road climbs a small hill, and after 1.3 km a bifurcation to your left will take you to Jalan Cahaya, the Hospital Road. After 3.1 km total distance, in front of the hospital, turn to the left and follow Jalan Padang Kerbau back northward along the Miri Hill. Pass Jalan Oil Well #1 for another 1.2 km and after 6.7 km total distance, turn left and drive up a small road for some 50 m to a parking area (Figure 3.9.1a).

3.9.2 Logistics and Safety

This visit to the quarry face situated in the back of a newly developed housing area takes about one hour. There are no specific dangers, but vigilance is required throughout the visit.

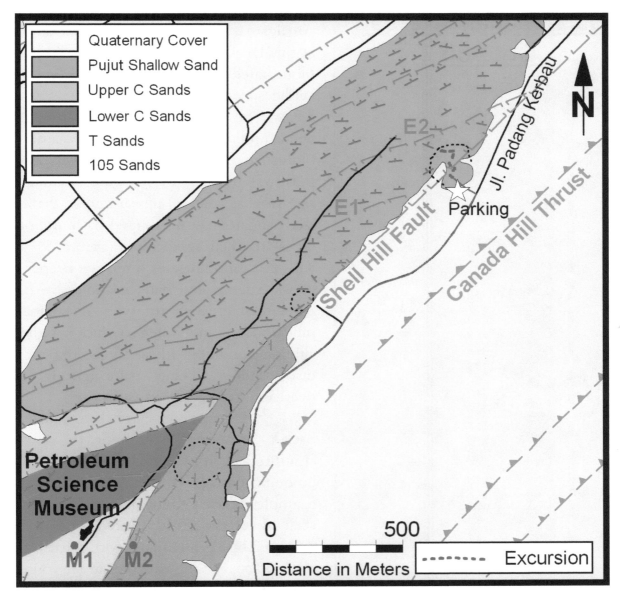

Figure 3.9.1a: Location and geological map of the Jalan Padang Kerbau Outcrop 2 (M1 and M2: location of wells Miri#1 and Miri#2)

3.9.3 Outcrop Highlights

- Exposures of the Pujut Shallow Sands (Figure 3.9.3a)
- The outcrop shows high Net-to-Gross sandstone content, characteristic of the better reservoir sequences in the field.
- Evidence for faulting during sediment deposition

3.9.4 Geological Description

From the parking area, follow the road and walk up along the hillside towards a newly-built area in a northerly direction. According to Schumacher's geological map, this outcrop

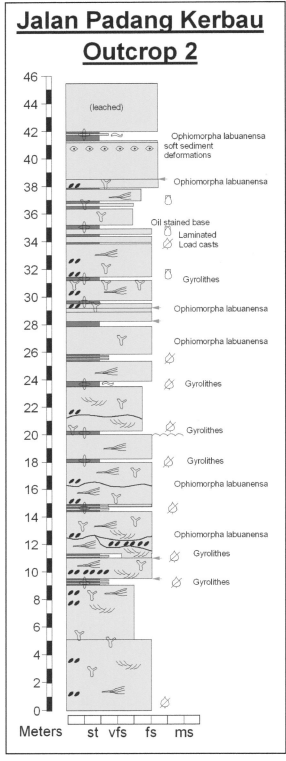

Jalan Padang Kerbau Outcrop 2

Figure 3.9.3a: Stratigraphic log of the "Pujut Shallow Sands", Jalan Padang Kerbau Outcrop 2

exposes parts of the Pujut Shallow Sand. Amalgamated sandstones are encountered at this level, partly cross-bedded and partly bioturbated. Shear zones (?) with limonitic surfaces dissect the sandstone (Figure 3.9.4a).The road leads to a wide path that winds up the old quarry, crossing a series of bioturbated sandstones (*Ophiomorpha* burrows) overlain by trough-cross bedded sandstones, passing upward into channelized sandstones, and topped by hummocky cross-stratified sandstones (Figure 3.9.4b). This last unit indicates a relative deepening in waterdepth and may be indicative of a transgressive event. Intervals of laminated mudstones and heterolithics indicate subtidal conditions; these levels are typically discontinuous, being truncated by the overlying, cross-bedded sandstones.

After a short climb, the path turns back, exposing hummocky cross-stratification, in a sandstone unit crossed by limonitic joints and minor faults. Some good examples of syn-sedimentary faulting can be seen along the path; a sandstone bed is dissected by a growth fault with an offset of some 30 cm but the fault dies in the younger sandstones immediately overlying the sandstone (Figures 3.9.4c). These structural instabilities during deposition of the Miri Formation are likely surface manifestations of the flow and lateral withdrawal of the mobile Setap shales at depth (Figure 3.8.4e).

The top of the quarry is reached after crossing bioturbated sandstones with minor and discontinuous mudstone intercalations. The area offers excellent views of Lambir Hill, and in clear weather, of the Mulu Mountains. Follow the same path to return to the car.

Note that Schumacher's geological map (Figure 3.9.1a) indicates that the Shell Hill Fault apparently crosses this outcrop, where it offsets the Pujut Shallow Sand against itself (implying a significantly smaller displacement compared with Jalan Padang

Kerbau Outcrop 1). However, there is no indication in the outcrop that the Shell Hill Fault should be present at this location.

Figure 3.9.4a: Shear zones (?) with limonitic surfaces dissect laminated and cross-bedded sandstones, "Pujut Shallow Sands", Jalan Padang Kerbau Outcrop 2

Figure 3.9.4b: Hummocky cross-stratification capping stacked channelized shoreface sandstones, "Pujut Shallow Sands", Jalan Padang Kerbau Outcrop 2

Figure 3.9.4c: Growth fault, "Pujut Shallow Sands", Jalan Padang Kerbau Outcrop 2

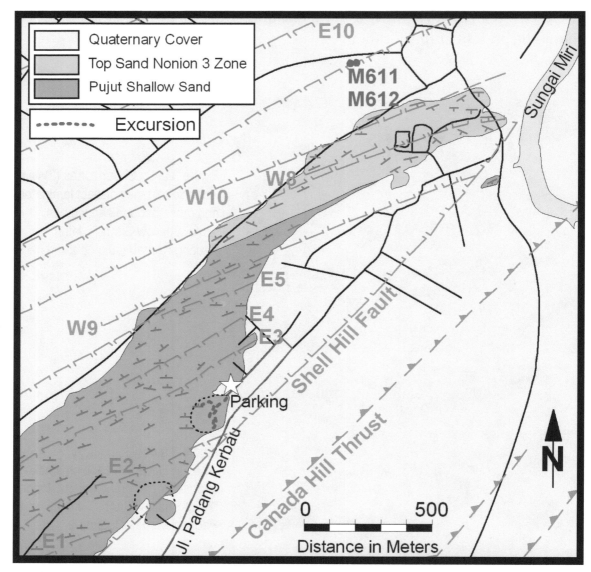

Figure 3.10.1a: Location and geological map of the Jalan Padang Kerbau Outcrop 3 (M611 and M612: location of wells Miri#611 and Miri#612)

3.10 Jalan Padang Kerbau Outcrop 3, near Lot 449

3.10.1 Access and Location

From the Clock Tower roundabout at km 0.0, follow Jalan Miri Bintulu southward (towards the airport). The road climbs a small hill, and after 1.3 km a bifurcation to your left will take you to Jalan Cahaya, the Hospital Road. After 3.1 km total distance, at the roundabout in front of the hospital, turn left and follow Jalan Padang Kerbau back northward along the Miri Hill. Pass Jalan Oil Well #1 and after 7.7 km total distance, turn left on Jalan Padang Kerbau 2. Turn left again and park at the end of the road, in front of Lot 435m (Figure 3.10.1a). The access to the old quarry is through a private property. At Lot 435, ask for the owner's permission to cross his backyard into the abandoned quarry area.

Jalan Padang Kerbau Outcrop 3

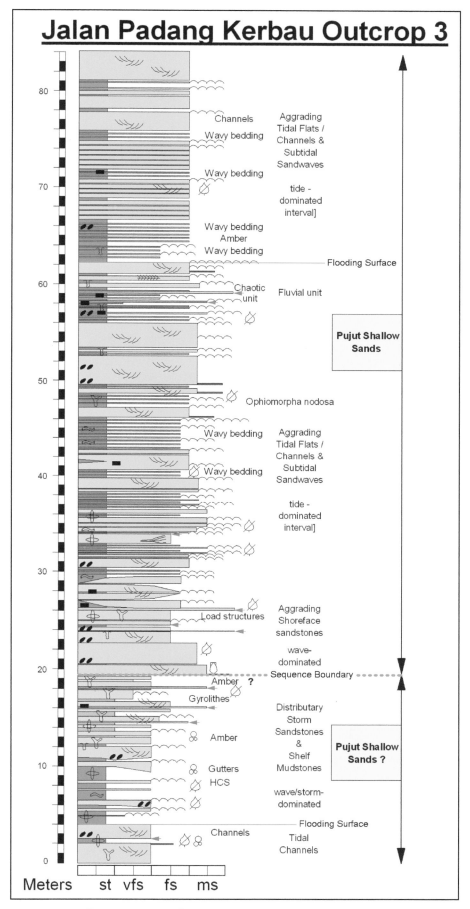

Figure 3.10.3a: Stratigraphic log of the "Pujut Shallow Sands", Jalan Padang Kerbau Outcrop 3

123

3.10.2 Logistics and Safety

The visit to this large, abandoned quarry takes one to two hours. Part of the quarry floor is swampy and a small fishpond has been created. Do not get close to the water, or to high cliffs. Otherwise, the visit does not present any particular danger.

3.10.3 Outcrop Highlights

- A spectacular section over 80 m thick, exposing a variety of sedimentary bodies ranging from offshore shelf to tidal environments (Figure 3.10.3a)
- Exposure of a lower depositional sequence characterized by a transgressive-regressive pattern, and an upper depositional sequence, characterized by an aggrading pattern

3.10.4 Geological Description

Upon entering the quarry, continue to your left, until you reach the base of the outcrop, opposite Lot 449 on Jalan Padang Kerbau. According to Schumacher's geological map, this outcrop exposes parts of the Pujut Shallow Sand. A 2 m high rock-face exposes a larger channel (1.5 m high and 8 m wide) with a smooth concave base (Figure 3.10.4a). The channel is filled with cross-bedded sandstone and includes *Ophiomorpha* burrows. This channel down-cuts an older channel system, which has incised a sequence of heterolithics along a recurring basal contact (Figure 3.10.4b). This outcrop provides an excellent vertical exposure of tidal channels; the upper distributary channel can be traced laterally, following the top of the small terrace in the back of the outcrop.

Figure 3.10.4a: Larger channel in tidal to shoreface sandstones, "Pujut Shallow Sands", Jalan Padang Kerbau Outcrop 3

124

Return towards the quarry, following the edge of the terrace. Some 20 m further, a cut perpendicular to the first outcrop exposes again a number of stacked channels of varying dimensions (Figures 3.10.4c,d). The top of the sandstones can be traced towards the hill, where a sharp contact with the overlying mudstones is exposed (Figure 3.10.4e). This contact represents a marine flooding surface and indicates an abrupt deepening of the depositional environments, with offshore mudstones overlying the tidal sandstones.

Figure 3.10.4b: Sandy, high energy system (tidal to shoreface bars) migrating laterally over low-energy tidal environment, "Pujut Shallow Sands", Jalan Padang Kerbau Outcrop 3

Figure 3.10.4c: Amalgamated, sandy tidal channels, "Pujut Shallow Sands", Jalan Padang Kerbau Outcrop 3

Figure 3.10.4d: Smaller tidal gully with clay clasts at the base, "Pujut Shallow Sands", Jalan Padang Kerbau Outcrop 3

Figure 3.10.4e: An overview of the "Pujut Shallow Sands", Jalan Padang Kerbau Outcrop 3: The lower sandy interval (left-hand side) is overlain by a mudstone-dominated section; the contact corresponds to an abrupt deepening of depositional environments

Figure 3.10.4f: Coarsening-upward trend, "Pujut Shallow Sands", Jalan Padang Kerbau Outcrop 3

Figure 3.10.4g: Strong currents at seabed created the hummocky cross-stratification in the sandstone bed, "Pujut Shallow Sands", Jalan Padang Kerbau Outcrop 3

Enter the main quarry to continue your field visit. From a distance, the stratigraphical succession is well exposed, as the strata dip towards Canada Hill. An overall coarsening upward sequence can be recognized, starting with mudstones, followed by a series of mudstones and interbedded sandstones, which are, in turn, overlain by stacked sandstones (Figure 3.10.4f).

Follow the edge of the quarry, crossing the mudstone sequences. The sandstones beds are initially thin (cm-dm) and consist of muddy sandstones. Continuing along the exposure, progressively more proximal sedimentary facies are encountered;

the sandstones become thicker and cleaner, and include hummocky cross-stratification (Figure 3.10.4g). Some beds thin out laterally, or are abruptly terminated (Figure 3.10.4h); excellent examples of gutter sandstones are encountered (Figure 3.10.4i). Occurring within an overall low-energy environment characterized by mudstone deposition, the sedimentary features in the sandstones are indicative of high-energy events, such as could be generated by severe storms. In continuation, the section exposes stacked sandstones, sometimes with a coarser grain-size; the amount of bioturbation increases and some trace fossils *(Gyrolithes)* indicate a return to tidal conditions. This overall regressive pattern is typical of highstand deposition.

Figure 3.10.4h: Abrupt termination of a sandstone bed along a gully, "Pujut Shallow Sands", Jalan Padang Kerbau Outcrop 3

Figure 3.10.4i: Gutter channel at the base of a sandstone bed, "Pujut Shallow Sands", Jalan Padang Kerbau Outcrop 3

A 5-6 m thick unit of stacked sandstones abruptly overlies this sequence. The lower contact is erosional and includes in places a lag of marine bivalves; the lowermost sandstone is medium-grained, and shows cross bedded stratification. This contact possibly represents a sequence boundary, as the overlying sediments (more than 60 m thick) are part of an aggrading system, an indication of a distinct stratigraphic architecture (Figure 3.10.3a).

Figure 3.10.4j: View from the back of the quarry, "Pujut Shallow Sands", Jalan Padang Kerbau Outcrop 3

Figure 3.10.4k: Stacked, eroded tidal bars overlain by tidal mudstones, "Pujut Shallow Sands", Jalan Padang Kerbau Outcrop 3

The overlying sequence is best observed on the northeastern side of the quarry (Figure 3.10.4j), which is reached by returning to the entrance of the quarry and turning left. The base of the section exposures a series of tidal channels characterized by wedge shaped sandstones, separated by mudstone drapes (Figure 3.10.4k). The interface between the mudstones and the sandstones shows occasional load structures, a type of soft-sediment deformation where the water-saturated muds were in places depressed by the weight of the overlying, soft sands (Figure 3.10.4l). Asymmetrical, current ripples can be observed on the surface of some sandstone beds (Figure 3.10.4m).

To continue the excursion, walk up the face of the quarry. This section of the quarry is characterized by shallow marine, tidal depositional environments consisting of sandstone units with rippled tops, draped by mudstones. The concave laminations and lateral cuts observed in the sandstone beds (Figure 3.10.4n) indicate deposition under high energy conditions such as ebb currents; the ripples at the top of the sandstones are the marks of the waves while the mudstone drapes indicate deposition under low energy conditions (tide cycle). Sandy tidal bars and shallow channels stack together to form beds of rather uniform thickness (Figure 3.10.4o). Other sandstone beds are coarse-grained, poorly sorted and disorganized (Figure 3.10.4p), a possible indication of a fluvial origin.

Figure 3.10.4l: Load structures along the mudstone-sandstone interface, "Pujut Shallow Sands", Jalan Padang Kerbau Outcrop 3

Figure 3.10.4m: Current ripples at top of sandstone bed, "Pujut Shallow Sands", Jalan Padang Kerbau Outcrop 3

Figure 3.10.4n: The cyclic alternation of high-energy (cross-bedded sandstones) and low-energy conditions (draping shales) is characteristic of tidal environments, "Pujut Shallow Sands", Jalan Padang Kerbau Outcrop 3

Figure 3.10.4o: Sandy tidal bar cannibalized by multiple tidal channels of small dimensions, "Pujut Shallow Sands", Jalan Padang Kerbau Outcrop 3

Figure 3.10.4p: Poorly sorted coarser grained sandstone bed, "Pujut Shallow Sands", Jalan Padang Kerbau Outcrop 3

Figure 3.10.4q: Laminated heterolithics abruptly overlay rippled sandstone beds, indicating a return to low-energy conditions in a tidal environment, "Pujut Shallow Sands", Jalan Padang Kerbau Outcrop 3

Figure 3.10.4r: Tidal bar sandstones with minor shale drapes, topped by a rippled surface, "Pujut Shallow Sands", Jalan Padang Kerbau Outcrop 3

Half way up the section, a thicker unit of cross-bedded sandstones containing thin mudstone drapes is overlain by a meter-thick sequence of laminated heterolithics (Figure 3.10.4q). Because of the contrast between the hard and the soft lithology, this contact is well exposed over a larger surface (Figure 3.10.4r). Here, the bed surface shows a higher order undulation (meter to multi-meter amplitude) created by strong waves; subsequent lowering in the water energy resulted in the formation of the ripples, which characterize the top of the sandstone.

The higher part of the quarry exposes thinner, rippled-top sandstones and laminated heterolithics (Figure 3.10.4s), including intervals with wavy bedding and soft sediment deformations such as flame structures (Figure 3.10.4t), all indicative of continuous tidal conditions.

The excursion continues to the uppermost terrace, at the base of the ultimate quarry face, consisting of stacked, cross-bedded sandstones. Return to your vehicle following the same path.

Figure 3.10.4s: Cyclic series of laminated sand-shale thin beds, representing high- and low-energy conditions, typical of tidal environments, "Pujut Shallow Sands", Jalan Padang Kerbau Outcrop 3

Figure 3.10.4t: Soft sediment deformation. Shortly after deposition, soft shales cut through the overlying sandstone bed and create "flame structures" (above pen, upper third of the photograph), "Pujut Shallow Sands", Jalan Padang Kerbau Outcrop 3

3.11 Temporary Holocene Exposures

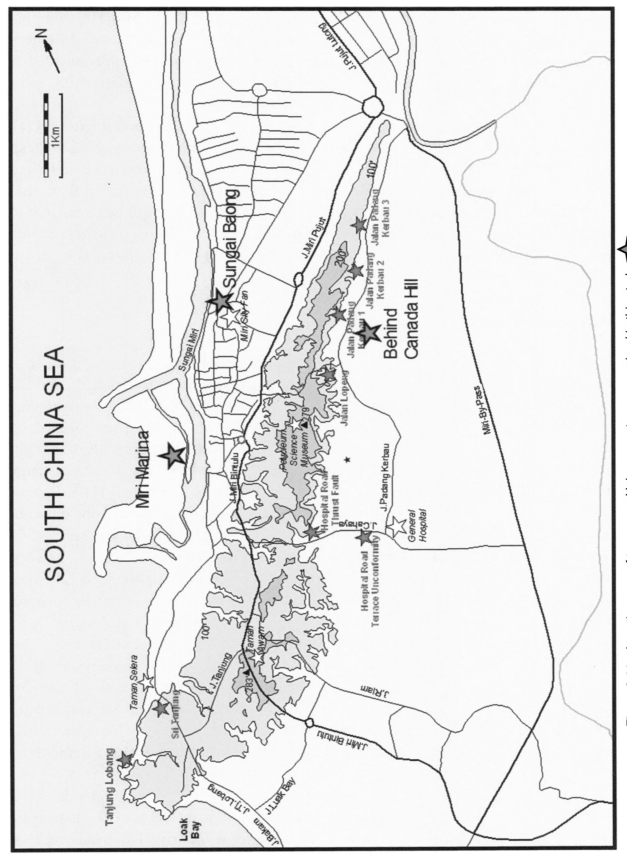

Figure 3.11a: Location map of temporary Holocene outcrops examined in this study

Miri is a fast growing town, and in many places excavations take place for building works. Wherever soil is being removed, opportunities exist for studying the shallow underground of the town. Aside from the elevated areas (Canada Hill, Miri Hill, Tanjong Lobang) these construction sites may provide information on the most recent geological history of the area. Generally the best opportunities are the deeper pits, especially if they are kept dry, so that sediments several meters thick may be visible. Regrettably, most excavations occur without any geologist studying the rocks.

In this section a few outcrops of Holocene age are described (Figure 3.11a). Although none of them allowed a detailed study of the sediments, they yielded interesting fossil faunas, which combined with information from other sources provide insight into the evolution of the Baram delta during the last 6000 years. It is hoped that other geologists working in the area, whether professionals or amateurs, will be stimulated to study the Holocene outcrops whenever these are exposed. The author (HR) is currently preparing a more detailed description of the mollusk faunas from the Holocene excavation sites, which will be published elsewhere.

3.11.1 Behind Canada Hill

In 1994-1995 a ditch near Jalan Lopeng was being deepened (Fig. 3.11a). Sand was dug up with a dragline, and deposited on the elevated dirt road parallel to the ditch, and the lower area in between the road and the ditch. As the dragline progressed, the sediments and fossils displayed their variations. Nearest to the hill the sediments consisted of boulders, with small numbers of shells that live on or bore into rocks (Plate 3.11.1a, Figures 1-4). This sand unit passed into fine sands with a typical tidal flat fauna (large numbers of specimens of a relatively low number of species, e.g. Plate 3.11.1a, Figures 9, 11, 13, 19; Plate 3.11.1b, Figures 1-3, 7, 8) and a few shells from an estuarine environment (brackish water species, e.g. Plate 3.11.1a, Figure 10; Plate 3.11.1b, Figure 11). Still further away the sediments passed into a mixture of coral sand and corals. As seen typically in shallow-water coral reefs the fauna included a large variety of coral types: solitary corals , branching corals, boulder-shaped corals, some small but others forming boulders up to a meter or more (Plate 3.11.1b, Figures 12-16; Plate 3.11.1c, Figures 1-13). These coral reef sediments contained a large number of mollusk species (Plate 3.11.1a, Figures 5-8, 12, 14, 15-17; Plate 3.11.1b, Figures 3, 6, 9-10), many of which presently do not occur in Sarawak or Brunei, albeit most are present in the coastal areas of western Sabah (from Labuan Island northward). Both the tidal flat and coral reef fauna comprise many closed doublets of bivalves, indicating that they lived in this very location. Beside corals and mollusks, numerous other fossils are present: Calcareous algae, foraminifera, sponges, bryozoans, worms, sea urchins, crabs and fish.

No direct dating method has been carried out so far (Carbon 14 dating is planned), but a comparison of our information with Caline & Huong (1961) suggests that these sediments were deposited around the time of the Holocene sealevel highstand, around 6,000 BP. In the Pleistocene, sealevel changed dramatically, and during

Plate 3.11.1a: Holocene gastropods from outcrops behind the Canada Hill

Figure 1: *Trochus maculatus* Linnaeus, 1758; Figure 2: *Turbo (Marmarostoma) brunneus* (Röding, 1798); Figure 3: *Astralium calcar* (Linnaeus, 1758); Figure 4: *Nerita (Ritena) chamaeleon* Linnaeus, 1758; Figure 5: *Polinices (Neverita) peselephanti* (Link, 1807); Figure 6: *Terebellum (Terebellum) terebellum* (Linnaeus, 1758); Figure 7: *Cypraea arabica* Linnaeus, 1758; Figure 8: *Cypraea staphylaea* Linnaeus, 1758; Figure 9: *Laevistrombus canarium* (Linnaeus, 1758); Figure 10: *Telescopium telescopium* (Linnaeus, 1758); Figure 11: *Murex trapa* Röding, 1798 with a specimen of *Crepidula (Siphopatella) walshi* Reeve, 1859 inside showing it was inhabited by a hermit crab; Figure 12: *Drupella rugosa* (Born, 1778); Figure 13: *Volema myristica* (Röding, 1798); Figure 14: *Pleuroploca trapezium* (Linnaeus, 1758) ; Figure 15: *Domiporta praestantissima* (Röding, 1798) ; Figure 16: *Vexillum (Costellaria) sanguisugum* (Linnaeus, 1758) ; Figure 17: *Vexillum (Vexillum) vulpeculum* (Linnaeus, 1758) Figure 18: *Architectonica perspectiva* (Linnaeus, 1758); Figure 19: *Atys (Atys) naucum* (Linnaeus, 1758)

Plate 3.11.1b: Bivalves and corals from Holocene outcrops behind the Canada Hill.
Bivalves: Figure 1: *Anadara (Tegillarca) granosa* (Linnaeus, 1758); Figure 2: *Anadara (Tegillarca) nodifera* (von Martens, 1860); Figure 3: *Vasticardium rugosum* (Lamarck, 1819); Figure 4: *Placuna (Ephippium) ephippium* Philipsson, 1788; Figure 5: *Isognomon (Isognomon) ephippium* (Linnaeus, 1758); Figure 6: *Macalia bruguierei* (Hanley, 1844); Figure 7: *Anomalocardia (Anomalodiscus) squamosa* (Linnaeus, 1758); Figure 8: *Marcia (Hemitapes) japonica* (Gmelin, 1791); Figure 9: *Placamen calophylla* (Philippi, 1836); Figure 10: *Circe (Circe) scripta* (Linnaeus, 1758); Figure 11: *Meretrix lusoria* (Röding, 1798).
Corals: Figure 12: *Symphyllia sp. (Mussidae)*; Figure 13: *Fungia fungites (Fungiidae)*; Figure 14: *Platygyra sp. (Faviidae)*; Figure 15: *Galaxea astreata (Oculinidae)*; Figure 16: *Acropora sp. (Acroporidae)*

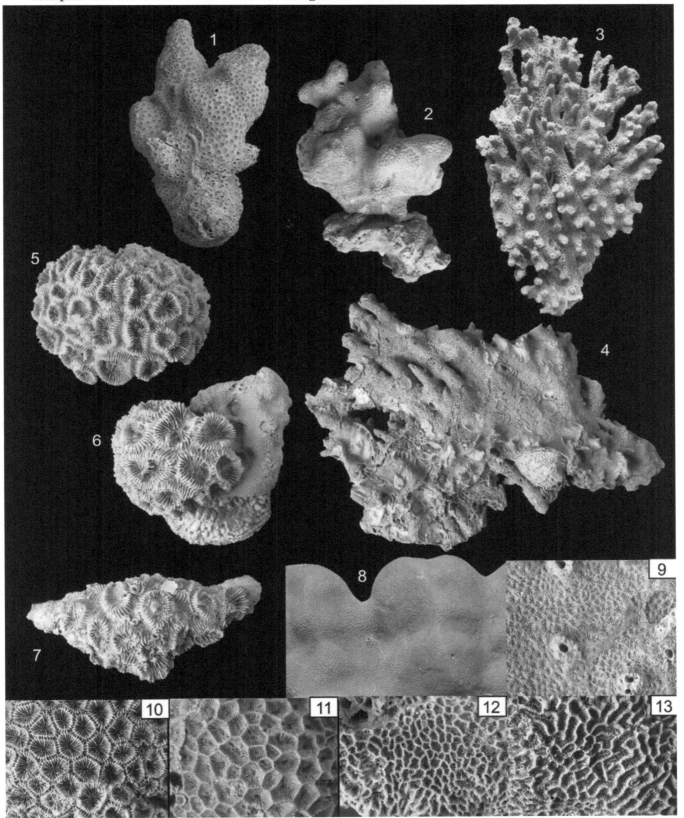

Plate 3.11.1c: Corals from Holocene outcrops behind the Canada Hill
Figure 1: *Cyphastrea sp. (Faviidae)* ; Figure 2: *Porites sp. (Poritidae)* ; Figures 3,4: *Acropora sp.*
(Acroporidae); Figures 5-7: *Favia sp. (Faviidae)*; Figure 8: *Porites sp. (Poritidae)*; Figure 9: *Pavona sp.*
(Agariicidae); Figure 10: *Montastrea sp. (Faviidae)*; Figure 11: *Favites sp. (Faviidae)*; Figures 12, 13:
Platygyra sp. (Faviidae).

the last glaciation (Weichselian) it dropped some 100-150 m below the present level. As a consequence, much of the Sarawak shelf (part of the Sunda shelf) was exposed, and the location where the Baram River discharged into the South China Sea shifted several tens of km to the northwest of the current river mouth. At the end of the last glaciation (about 10,000y BP) a rapid sealevel rise commenced, which around 8,000y to 6,000y BP drowned the Baram river valley and thus formed a shallow embayment in which coral reefs could establish, as is demonstrated by the fossils found at this temporary excavation site. This period probably lasted only a few hundred years, as the Baram River delivered so much sediment that the embayment quickly filled and the river delta evolved into its present form. Currently such shallow-water coral reefs are not found anywhere near the Baram delta. Small coral banks do occur in waters 5 to 40 m deep in the open sea, southwest of Tanjong Lobang and offshore Brunei, but there are no tidal flats behind them. The nearest area with similar conditions is Labuan Island, offshore Sabah.

3.11.2 Sungai Baong Outcrop

In 1996 a storm drain (Sungai Baong, Fig. 3.11a), not far from the municipal swimming pool) was repaired. Much mud was removed, and sand deposits were exposed down to several meters below the current sea level. The excavated sediments contained a rich fossil fauna dominated by mollusk shells, but also some corals, foraminifera, bryozoans, worms, sea urchins, barnacles, crabs, and fishes. The sediments consisted of fine, bluish sand and the fossils were colored bluish grey or salmon (in the case of thick-shelled gastropod fossils). Many of the bivalve shells were present as doublets in living position.

The fossils are diagnostic of different depositional environments. The fauna are dominated by shells from shallow subtidal sandy areas (foreshore deposits), but some shells have been washed in from salt marsh, beach and intertidal to shallow subtidal rocky areas. These deposits probably have a similar age as those found from the sites behind Canada Hill described above. They were deposited in a shallow open sea, a few hundred meters from the beach.

3.11.3 Miri Marina

Construction work of a new outlet for the Miri River was started in 1997. This resulted in the building of a marina over a large tract of reclaimed land near the city centre. Dikes were built from large blocks, and the area in between was filled with sand. This sand was dredged a few kilometers from the shore, and proved to be very rich in seashells. These shells are not fresh, but preserve their original color pattern; they all probably originated from Holocene deposits (Carbon-14 dating is in progress).

The fauna are very rich in several species, including some which are not known to presently live in the sea near Miri. Sediments with a similar fauna occur along a large stretch of the coast and were also used to enlarge the beach at Jerudong, Brunei.

3.11.4 Other Outcrops

Aside from outcrops, Holocene fossils can be found on most beaches in North Sarawak and Brunei, as the sea is continuously eroding the Holocene sediments. These fossils can generally be recognized based on their preservation, especially their colors – which are similar to those found in the Sungai Baong outcrop or the sediments at Miri Marina.

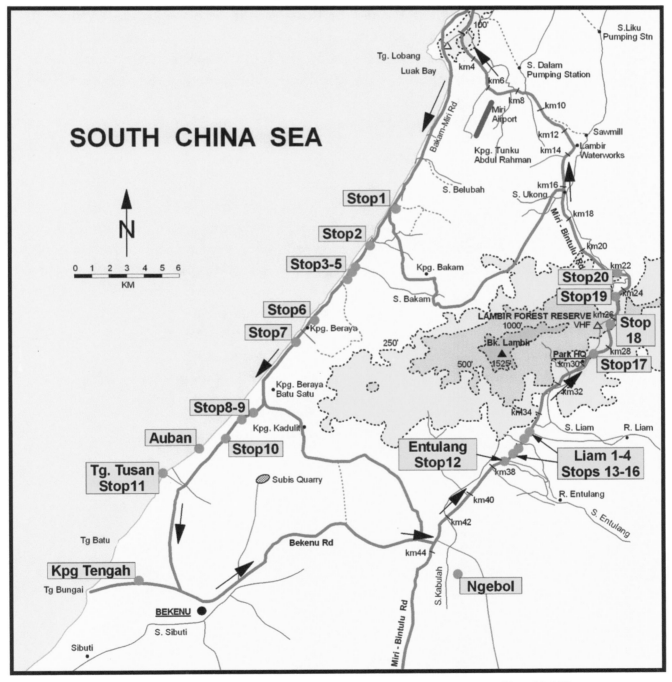

Figure 3.12.1a: Route map and location of stops for the field trip around Lambir Hills

3.12 Tour around Lambir Hills

3.12.1 Access and Location

The trip is a counter-clockwise visit of the Lambir Hills, following the asphalted Miri-Bekenu road and returning on the Miri-Bintulu road. The total length of the tour is slightly over 110 km.

From the Clock Tower roundabout, km 0.0 in Miri, make your way southeastward towards Taman Selera Park, pass Tanjung Lobang and continue on the main road to Bekenu. For the first 20 km, the asphalted road crosses a lowland area, in close vicinity to the beach; it then continues through terraces and low hills before reaching the plain of Bekenu. At the entrance of Bekenu, after some 50 km, turn left and continue driving on the main road till reaching the junction with the Miri-Bintulu road, another 18 km journey. Turn left to return to Miri (Figure 3.12.1a).

The detailed access and location and side tours are described below.

3.12.2 Logistics and Safety

This is a day-long excursion, mostly along asphalted roads which at times experience heavy traffic. Great care to be taken in choosing safe parking locations and in crossing the road. Visits to the cliff areas and mud volcanoes require additional care in keeping a safe distance from potentially hazardous areas, such as cliff edges and mud pools. It is best to go in a group, with several drivers who can take turns at the wheel; drivers should be careful to always watch the road and the traffic and not be focusing on the geology! The vehicle should be in good condition and the tank full of gasoline before starting the trip. Take plenty of water and light snacks; a hot meal can be enjoyed in Bekenu.

For more information on a visit to the Lambir Hills National Park, please consult the Sarawak Forestry Department's website at:
http://www.forestry.sarawak.gov.my/forweb/np/np/lambir.htm
Further detailed descriptions of the park, its flora and fauna can be found in Hazebroek and Abang Morshidi (2000).

3.12.3 Outcrops Highlights

- Sibuti, Lambir, Miri and Tukau formations (Figure 3.12.3a)
- Terrace unconformities
- Tidal sediments and soft-sediment deformations
- Spectacular sea cliffs at Tusan

3.12.4 Geological Description, Miri-Bekenu

The first outcrop (Stop 1) is reached after some 18.7 km of driving, at the clearing site of a

GEOLOGICAL MAP, LAMBIR HILLS

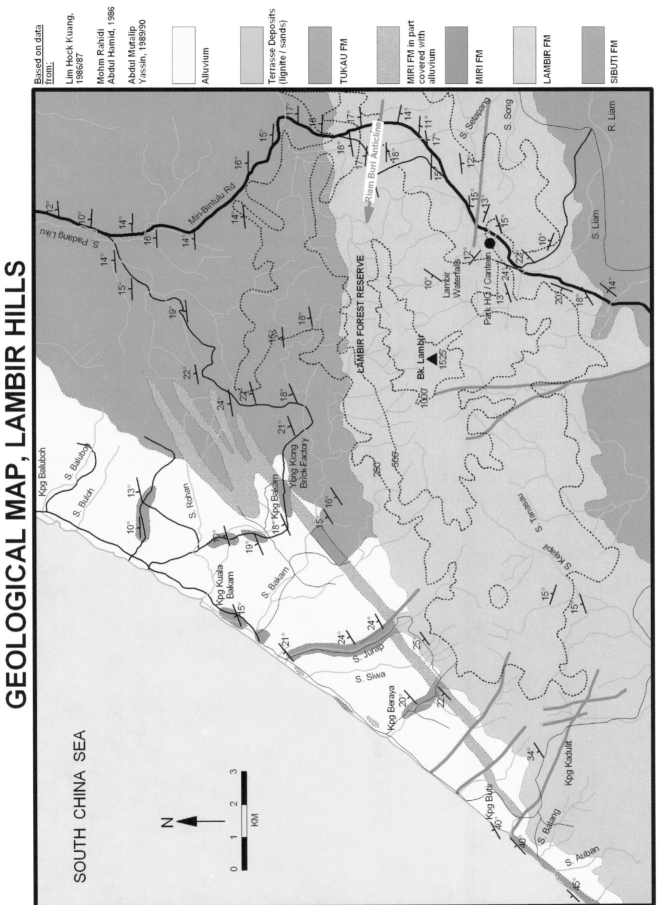

Figure 3.12.3a: Geologic Map of Lambir Hills

143

new housing area, on the left-hand side of the road, where a low amplitude anticlinal structure can be viewed in the hillcut (Figure 3.12.4a). Terrace deposits truncate these Mid Miocene sediments of the Miri Formation (Figure 3.12.4b), a feature that will be seen in detail in some of the following outcrops. Oil seeps have been recorded here, within the Miri Formation.

After passing the left junction to Kampong Bakam, the road winds up a low hill and terrace area. Stop 2 is located at km 22.8 (N 04° 13.205', E 113° 54.247'), at the base of a low cliff, on the right-hand side of the road (Figure 3.12.4c). This outcrop exposes a number of tidal channels cut into bioturbated sandstones and backfilled with mudstone (Figure 3.12.4d); coal clasts and quartz granules indicate the proximity to the coastline. Note the smell of petroleum coming from the rocks! Shortly after this outcrop, a bifurcation at km 23.6 leads to Beraya Beach, known for the presence of wave transported fossil crabs and other marine fossils (see Chapter 4.4). Beraya Beach is also the starting point of the Beraya-Tusan Beach drive.

Continue until km 23.2 (Stop 3; N 04° 13.004', E 113° 54.094') where another mudstone-filled channel is exposed on the right-hand side of the road (Figure 3.12.4e). The channel has incised poorly-sorted, medium-grained sandstone with quartz granules; flaser wavy bedding and lenses of brownish clays indicate alternate high and low energy conditions, including possible subaerial exposure within the Miri Formation.

Some 700 m further down the road (Stop 4 at km 23.9) stop to observe another mudstone-filled channel (Figure 3.12.4f). This outcrop of the Miri Formation exposes a sandstone bed whose surface is indented by a series of decimeter-scale

Figure 3.12.4a: Folded Miri Formation truncated by terrace deposits, Bakam Road Junction

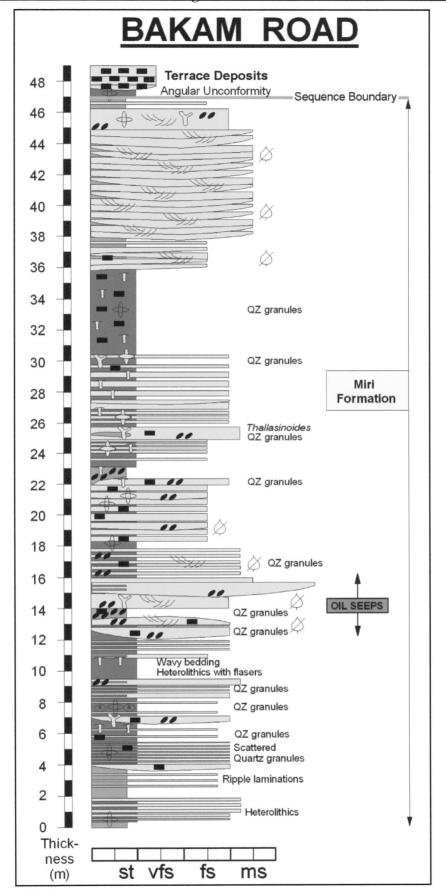

Figure 3.12.4b: Stratigraphic log of the Miri Formation, Bakam Road Junction

gullies, backfilled with mudstone (Figures 3.12.4g).

At km 26.6, the road passes along a low trench where the grayish-blue beds of the Miri Formation are cut at a low angle by terrace deposits (Stop 5, Figure 3.12.4h). These terrace deposits consist of through cross-bedded (Figure 3.12.4i) and low-angle cross-laminated, brownish sandstones, overlain by leached, white sandstones (informally the "lower" and "upper terrace sandstones).

Shortly after this outcrop, a series of fossiliferous beds of the Miri Formation are exposed along the road between kilometers 27.5 to 30.8, with beautifully preserved fossil crabs and mollusks (see Chapter 4.5).

Figure 3.12.4c: Outcrop overview, Miri Formation, Stop 2

Figure 3.12.4d: Mudstone-filled tidal channels, Miri Formation, at Stop 2

146

Figure 3.12.4e: Mudstone-filled tidal channel, Miri Formation, at Stop 3

Figure 3.12.4f: Mudstone-filled tidal channel, Miri Formation, at Stop 4

Figure 3.12.4g: Small mudstone-filled gully, Miri Formation, Stop 4

Figure 3.12.4h: Angular unconformity at Stop 5: Vertical beds of the Miri Formation overlain by horizontal Pleistocene sandstones (Terrace Unconformity along dashed line)

Figure 3.12.4i: Trough cross-bedded sandstones in Pleistocene "lower" terrace
deposits, Stop 5

Immediately after passing Sungai Uban bridge (Stop 6 at km 30.7), the unconformity at the base of the terrace deposits is particularly well exposed on the left-hand
side of the road. At this location, beds of the Miri Formation are dipping steeply, and
the angularity of the contact with the overlying terrace deposits is very obvious (Figure
3.12.4j). This unconformity is further enhanced by the color contrast between the
grayish-blue Miocene sediments below and the brown and white Pleistocene terrace
deposits above. Poorly-sorted sandstones with quartz granules rest on the truncated
marine sandstones of the Miri Formation (Figure 3.12.4k); they are, in turn, overlain by
brownish trough-cross bedded sandstones (Figure 3.12.4l) and leached white
sandstones, as also observed in the previous Stop (see also Chapter 3.6). The identical
stratigraphic composition of these terrace deposits is a strong indication for a similar
age; Wilford (1961) assigns a Pleistocene age to these terrace deposits.

This Terrace Unconformity is observed further along the road; at km 34.5
(Stop 7), spectacular outcrops of the unconformable contact can be seen on both sides
of the road (Figure 3.12.4m).

Drive further until km 35.6 (Stop 8; N 04° 07.717', E 113° 50.106') where a
deep cut on the hillside reveals a number of antithetic normal faults (Figure 3.12.4n).
The faulted sequence of the Miri Formation has a high sand content, yet all fault traces
are lined with a clay smear (Figure 3.12.4o). These faults possibly correspond to early
extensional normal faults rotated by the subsequent structural compression (M.

Wiemer, personal communication), as the bedding dip (80-85°) and occasional slickensides indicate a low angle oblique movement. The freshness of fault features suggests late lateral movement of Riedel shear type, which would correspond to the late faulting at right angles to the coast (see Chapter 4.5.4).

Figure 3.12.4j: Angular unconformity between steeply dipping beds of the Miri Formation and sub-horizontal Pleistocene sandstones, Stop 6 (Terrace Unconformity highlighted with white dashed line)

Figure 3.12.4k: Detail of the angular unconformity between steeply dipping beds of the Miri Formation and sub-horizontal Pleistocene sandstones, Stop 6 (Terrace Unconformity highlighted with white dashed line)

Figure 3.12.4l: Pleistocene trough cross-bedded sandstones, Stop 6 ("lower" terrace sandstones)

Figure 3.12.4m: Angular unconformity between steeply dipping beds of the Miri Formation and sub-horizontal Pleistocene sandstones, Stop 7 (Terrace Unconformity shown by the dashed line)

Figure 3.12.4n: Outcrop overview, faulted Miri Formation, Stop 8

Figure 3.12.4o: Clay smear along fault plane, Miri Formation, Stop 8

Continue for another kilometer and you will reach Stop 9, at km 36.0, where you can to observe a series of laterally migrating tidal channels on the right-hand side of the road (Figure 3.12.4p). This channel-fill architecture consists first in a drape by rhythmic heterolithics, followed by sandstone deposition (Figure 3.12.4q). Larger wave-forms shape the sandstones deposited under high-energy conditions. The alternation of low and high-energy conditions is also well illustrated by the superposition of cm-thick rhythmic heterolithics, and intervals characterized by abundance of clay clasts (Figure 3.12.4r). This outcrop of the Miri Formation also exposes textbook examples of load structures, created in waterlogged mudstones and sandstones shortly after their deposition. Figure 3.12.4s shows a sandstone bed segregated into a number of pillow-shaped bodies, each displaying a different stadium of internal rotation and being progressively embedded into the underlying soft mudstone. In places, the soft sands are injected into the overlying sediments, creating small internal folds (Figure 3.12.4t).

At a short distance on the right-hand side of the road (Stop 10 at km 38.8), a perpendicular quarry face allows further sedimentological observations of tidal deposits within the Miri Formation. A near continuum of lenticular, wavy and flaser bedding can be recognized here (Figure 3.12.4u).

Figure 3.12.4p: Stacked, laterally migrating tidal channels, Miri Formation, Stop 9

153

Figure 3.12.4q: Detailed view of stacked, laterally migrating tidal channels, Miri Formation, Stop 9 (downcutting channel base highlighted by dashed lines)

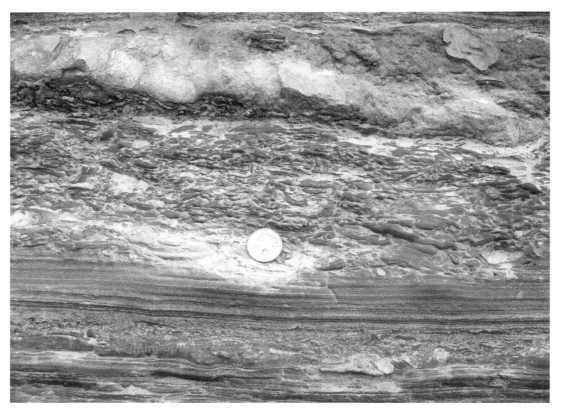

Figure 3.12.4r: Beds formed by clay clasts accumulations in tidal environments, Miri Formation, Stop 9

Figure 3.12.4s: Soft sediment deformations in a tidal depositional environment: Upward flowing mudstones partition and upturn overlying sandstone-shale fine layers, Miri Formation, Stop 9

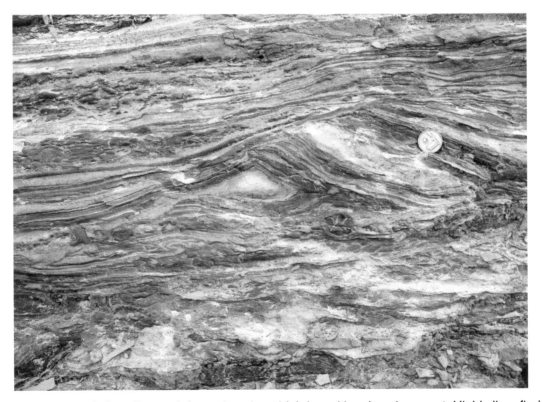

Figure 3.12.4t: Soft sediment deformations in a tidal depositional environment: Highly liquefied, mobile sand-mudstone units entirely reshape the original depositional stratigraphy along the base of a tidal channel, Miri Formation, Stop 9

Figure 3.12.4u: Laminated sand-mudstone couplet in a tidal unit, Miri Formation, Stop 10

Figure 3.12.4v: Large current-rippled surface, Lambir Formation, Tusan Beach, Stop 11

At km 40.9, a bifurcation to the right leads to Tusan Beach (Stop 11), a further 1.5 km distance. The parking lot at the end of the road offers a spectacular view of the high cliffs along coastline. Stacked sandstone deposits of the Lambir Formation dip towards the sea at an angle of 37°; a large bedding plane surface exposes assymetrical ripples reflecting seaward paleo-flow (Figure 3.12.4v). These Miocene sediments are truncated by a high Pleistocene terrace, where the leached white sandstones are seen outcropping (Figure 3.12.4w). At low tide, the 250 m thick succession of the Lambir Formation can be followed uninterruptedly. This exposure is characterized by a number of coarsening-upward parasequences, occasionally dissected by fault zones along which oil seeps are common (Figure 3.12.4x). Tusan Beach is also an alternate point of departure for a visit to the Beraya-Tusan Seacliffs, described separately (Chapter 4.6) and in Lesslar & Wannier (1998).

Continue in the direction of Bekenu, passing Kampong Terahad and Kampong Angus, with bifurcation at km 42.6 which leads through the Sibuti Formation to the Kampong Tengah fossil outcrop (see Chapter 4.7) and ends at Bungai Beach. Further down the road, a bifurcation at km 43.7 leads to the town of Bekenu. The road ahead takes you towards Sibuti, Batu Niah and Bintulu. This part of the excursion crosses and follows the West Baram Line, yet this major fault is not expressed as a prominent topographic feature, except that it separates the Lambir Hills from the flat plains with outcrops of the Setap shales.

Figure 3.12.4w: Angular unconformity between steeply dipping beds of the Lambir Formation and sub-horizontal Pleistocene terrace sandstones, Tusan Beach, Stop 11

Figure 3.12.4x: Oil seeps on the beach in front of a fault zone,
Lambir Formation, Tusan Beach, Stop 11

3.12.5 Geological Description, Bekenu-Miri

Keep to the left on the main road that joins with the Miri-Bintulu road after 17.9 km. At the junction, take left, and return in the direction of Miri. After 100 m, the Beluru junction is passed, being the access road to the Ngebol mud volcanoes (described in Chapter 3.13). Continue driving on the main road to Miri for another 7.5 km, and stop on the left-hand side of the road, on top of a hill, next to a high outcrop (N 04° 08.844', E 114° 00.390'). This location is known as Entulang (Stop 12, Figure 3.12.5a) and exposes some 45 m of Middle Miocene sediments, including the contact between the Sibuti and the overlying Lambir Formation (Simmons et al., 1999). This outcrop as well the following ones are described and illustrated in the CD Destination Miri (Lesslar and Wannier, 1998).

158

The Entulang outcrop exposes the fossiliferous, early Middle Miocene (Langhian) Sibuti Formation, dated by marker nannofossil species to belong to the biozone NN4/5. These marine deposits are open-shelf mudstones deposited in a distal offshore setting, and are characterized by a rich and diverse fossil assemblage. A maximum flooding surface can be recognized within this approximately 20 m thick unit (Figures 3.12.5b). An abrupt contact with the sandstones of the overlying Lambir Formation (consisting of wave-dominated shoreface and subtidal deposits) is interpreted to be a sequence boundary (Figure 3.12.5c) as it shows a sharp retreat of depositional environments. Higher up in the section, hummocky cross-stratification can be observed (Figure 3.12.5d). At Entulang, the Lambir Formation is correlated to the late Middle Miocene (Serravallian), and the basal sequence boundary likely corresponds to the Langhian-Serravallian boundary (13.8 Ma).

After following the road downhill for one kilometer, you will reach the Liam 1 outcrop (Stop 13, Figure 3.12.5e) which exposes a 15 m high hill cut on the left hand side of the road (Figure 3.12.5f). The contact between the Sibuti and the Lambir formations is again exposed here. Evidence for erosion can be seen at the base of the sandstones, where fossil leaf imprints (Figure 3.12.5g) can be found. The Lambir Formation sandstones show significant lateral facies variations and also expose various examples of soft sediment deformation features (Figure 3.12.5h), including slump folds and possible sandstone pipes. Enigmatic cylindrical sandstone bodies piercing the bedding vertically may represent escape features of pressurized, liquefied sands, created shortly after deposition (Figure 3.12.5i).

A number of outcrops are scattered along the road leading to the Lambir Park Headquarters, which are described in Lesslar & Wannier (1998); these outcrops represent successively younger strata of the Lambir Formation.

Liam 2 (Stop 14) is a cliff over 50 m high, situated on the left-hand side of the road, some 600 m past Liam 1 (Figure 3.12.5j). Stacked tidal channels (Figure 3.12.5k), and rippled surfaces (Figure 3.12.5l) are particularly well exposed here.

Figure 3.12.5a: Overview of the Entulang outcrop (MFS = Maximum Flooding Surface; SB = Sequence Boundary), Stop 12

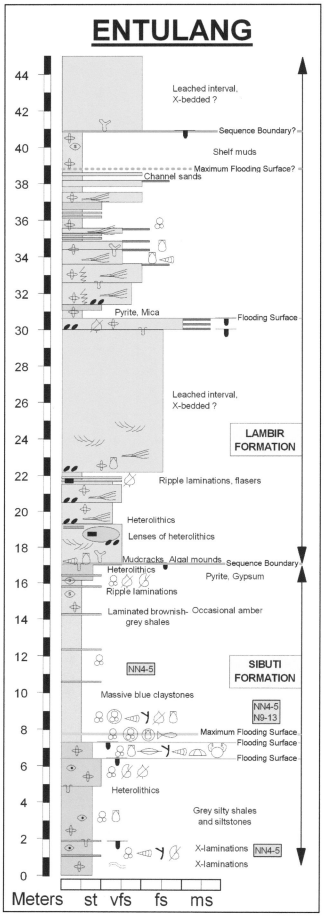

Figure 3.12.5b: Stratigraphic log of the Sibuti and Lambir Formations, Entulang outcrop, Stop 12

160

Figure 3.12.5c: Sequence boundary separating the Sibuti
Formation (below) from the Lambir Formation (above)

Figure 3.12.5d: Hummocky cross-stratification in shoreface sandstones of the Lambir
Formation. Entulang, Stop 12

Figure 3.12.5e: Overview of the Liam 1 outcrop across the road, Stop 13

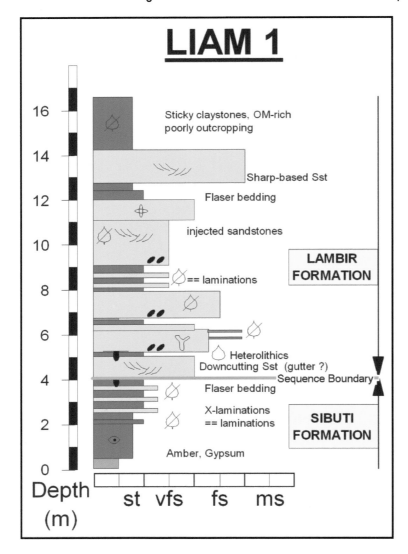

Figure 3.12.5f: Stratigraphic log of Sibuti and Lambir Formations, Liam 1, Stop 13

Figure 3.12.5g: Lambir Formation sandstone bed with fossil leaves, Liam 1, Stop 13

Figure 3.12.5h: Soft sediment deformations (block rotations and fold-forming mudstone flows) at the edge of a channel cut, Lambir Formation. Liam 1, Stop 13

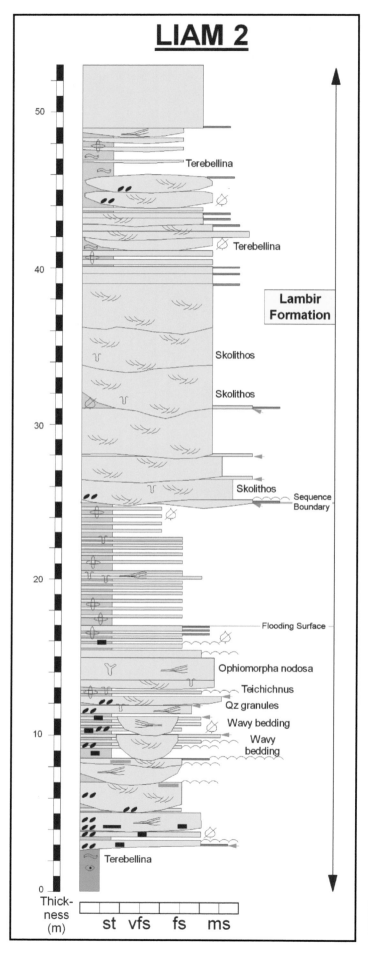

LIAM 2

Terebellina

Terebellina

Lambir Formation

Skolithos

Skolithos

Skolithos — Sequence Boundary

Flooding Surface

Ophiomorpha nodosa

Teichichnus

Qz granules

Wavy bedding

Wavy bedding

Terebellina

Thickness (m)

st   vfs   fs   ms

Figure 3.12.5j: Stratigraphic log of Lambir Formation, Liam 2, Stop 14

164

Figure 3.12.5i: Sandstone pipe in Lambir Formation, Liam 1, Stop 13

Figure 3.12.5k: Stacked tidal channels in Lambir Formation. Liam 2,
Stop 14

Figure 3.12.5l: Rippled surface at the top of a sandstone unit,
Lambir Formation. Liam 2, Stop 14

Liam 3 (Stop 15) outcrop is located on the left-hand side of the road, some 200 m further and exposes an intra-formational unconformity (Figure 3.12.5m), possibly related to sediment loading on deep unstable shales. Tidal sediments with soft sediment deformation and cut-and-fill structures (Figure 3.12.5n) can be observed here.

Some 300 m further, on the left-hand side of the road, the Liam 4 outcrop (Stop 16) exposes a series of tidal channels (Figure 3.12.5o), overlain by through cross-bedded shoreface sandstones (Figure 3.12.5p), and, in turn, covered by mudstones. A marine fossil fauna is present in the upper part of the mudstones; based on nannofossils (NN5 biozone), this mudstone interval is dated as Lower Serravallian, about 14 Ma (Figures 3.12.5q).

Approximately one kilometer after passing the Lambir National Park entrance, tidal sediments with internally down-cutting tidal channels can be seen on the left-hand side of the road (Lambir Formation, Stop 17, Figure 3.12.5r).

The road soon climbs the flank of the Riam Buri Anticline, where larger cuts expose the Lambir Formation on both sides of the road (Stop 18). An intra-formational angular unconformity and erosional truncation is seen shortly before reaching the top of the hill (Figure 3.12.5s). A major fault zone crosses the road underneath the VHF tower, (Figure 3.12.5t) and a larger channel-cut can be observed a short distance further, on the right-hand side of the road (Figure 3.12.5u).

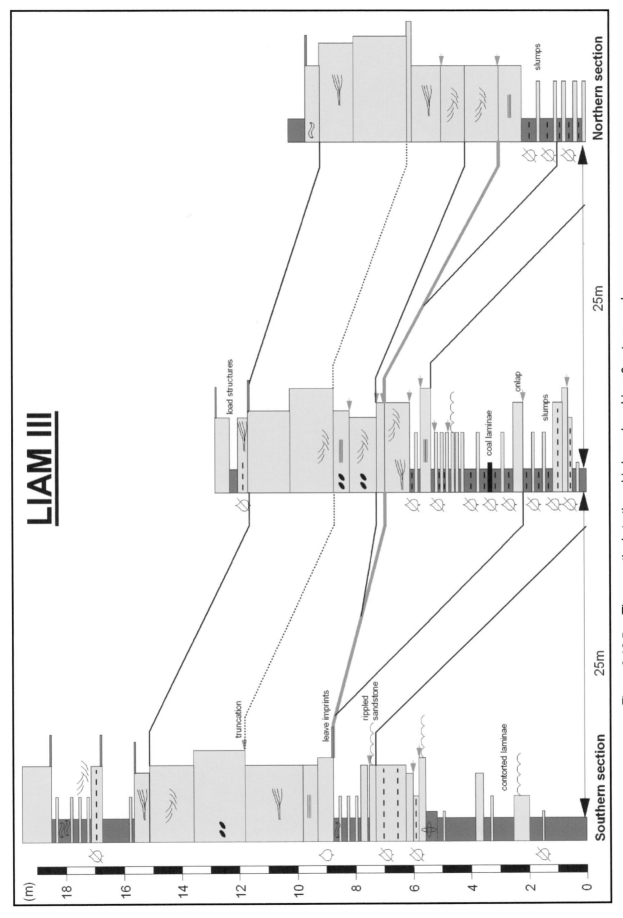

Figure 3.12.5m: Three vertical stratigraphic logs along Liam 3 outcrop and location of intra-formational unconformity within the Lambir Formation, Stop 15

Figure 3.12.5n: Lateral edge of a sandy tidal bar abutting against tidal bay
sand-mudstone couplets. Lambir Formation, Liam 3, Stop 15

Figure 3.12.5o: Low-amplitude tidal channels, Lambir Formation, Liam 4, Stop 16

Figure 3.12.5p: Trough cross-bedded shoreface
sandstones, Lambir Formation, Liam 4, Stop 16

Figure 3.12.5r: Down-cutting tidal channels, Lambir Formation, Stop 17

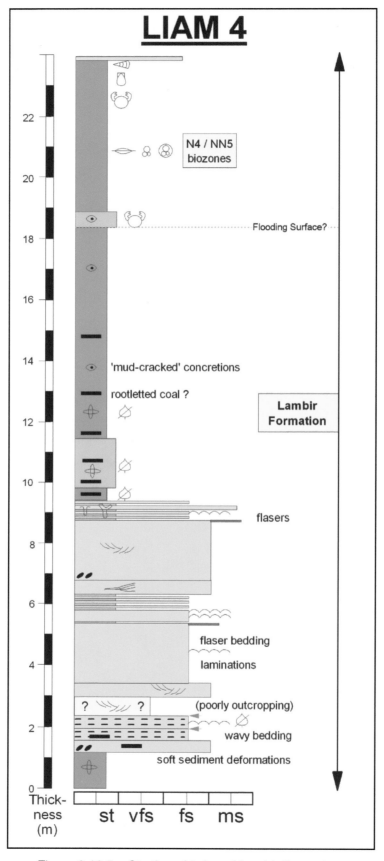

Figure 3.12.5q: Stratigraphic log of Lambir Formation,
Liam 4, Stop 16

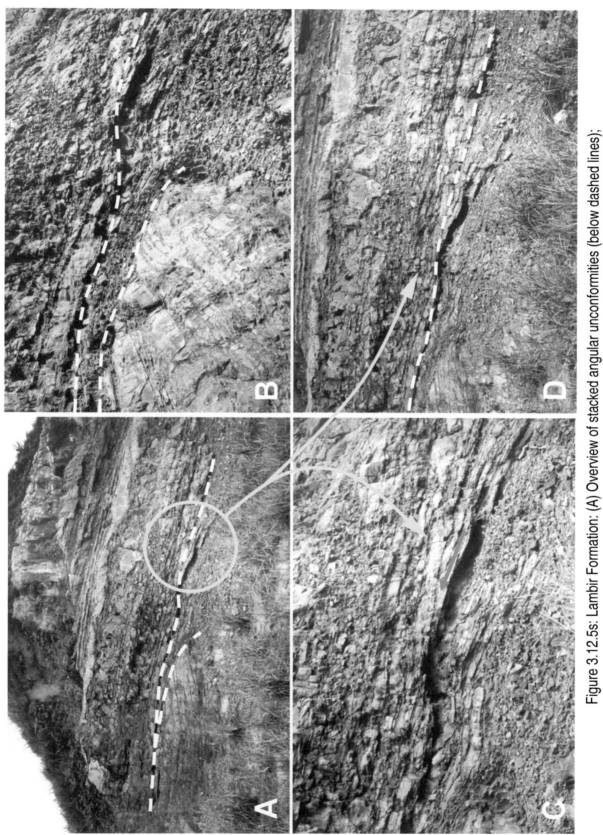

Figure 3.12.5s: Lambir Formation: (A) Overview of stacked angular unconformities (below dashed lines);
(B) Detailed view showing the high-angle unconformity at the base (lower dashed line) and the low-angle unconformity above (upper dashed line);
(C,D) Detailed view of the upper, low-angle truncation (red arrow below truncation in C and dashed white line along unconformity in D).

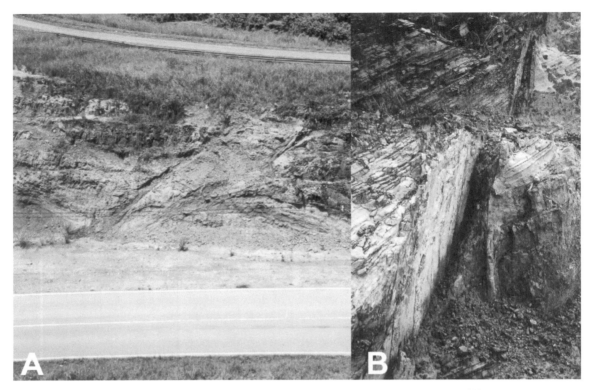

Figure 3.12.5t: Fault zone on top of the Riam-Buri anticline, Stop 18: (A): wide, low-angle fault zone to the north of the road; (B) narrow, shale-filled, high-angle fault zone to the south, across the road

Figure 3.12.5u: Larger-scale (distributary?) tidal channel, Lambir Formation, Stop 18 (dashed white line marks the base of the erosional channel)

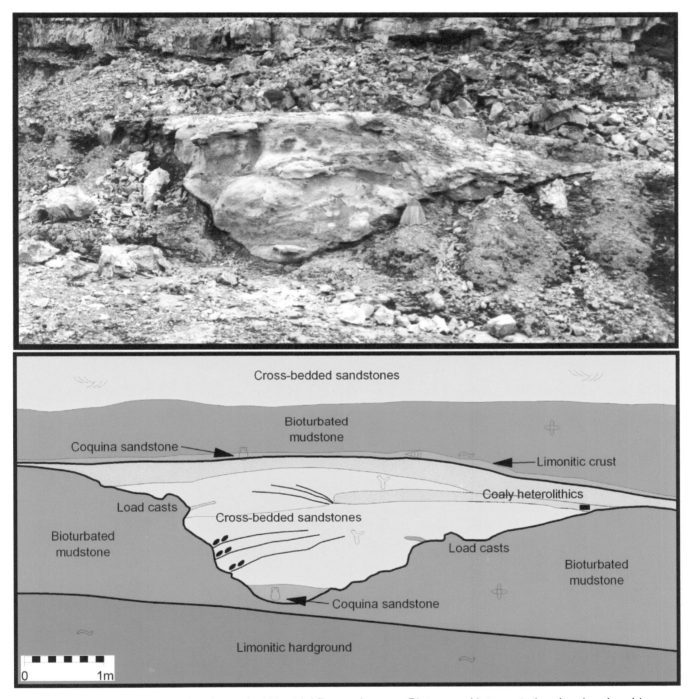

Figure 3.12.5v: Sandy tidal channel within tidal flat mudstones: Picture and interpretative drawing, Lambir Formation, Stop 19

After driving 85.6 kilometers (N 04° 13.938', E 114° 03.475'), as the road bends sharply to the right before descending towards the Liku valley, outcrops a 5 m-wide tidal channel cut into bioturbated tidal flat mudstones (Stop 19, Figure 3.12.5v). The bottom of this gutter-shaped channel is characterized by a shell lag, and load casts are visible on the flanks. The main body of the channel is composed of trough cross-bedded sandstones, in places with stacked layers of clay clasts, and *Ophiomorpha* burrows. The upper part of the channel consists of lower energy deposits, such as heterolithics, occasionally rich in disseminated organic matter. A thin fossiliferous level drapes the channel, representing a marine flooding event.

Some 2.3 km further (Stop 20), a small quarry on the right-hand side of the road exposes the Tukau Formation, with offshore bars showing lateral accretion; non-branching, sub-vertical tubular trace fossils *(Skolithos)* can be observed at this locality (Figure 3.12.5w).

For most of the return journey back to Miri, the road crosses the poorly consolidated sandstones of the Tukau Formation; Miri is at a distance of some 22.2 kilometers from the last outcrop (Stop 20).

Figure 3.12.5w: Stacked tidal bar sandstones with *Skolithos* trace fossils, Tukau Formation, Stop 20

3.13 Ngebol Mud Volcanoes and Bulak Setap-3 Wellhead

3.13.1 Access and Location

About 15 km southwest of the Lambir National Park Headquarters, along the Miri-Bintulu road, is the junction to Logan Bunut National Park and Beluru. Turn left at this junction and drive for 2.1 km to reach the start of the trail to Ngebol Mud Volcano. Park your car on right side of road and follow the path leading to Ngebol Mud Volcano. The 2 km hike (one-way) takes about 30-45 minutes walking through an oil palm plantation. From the car, the walk to the well head takes less than 10 minutes. See Figure 3.13.1a for a map with GPS coordinates.

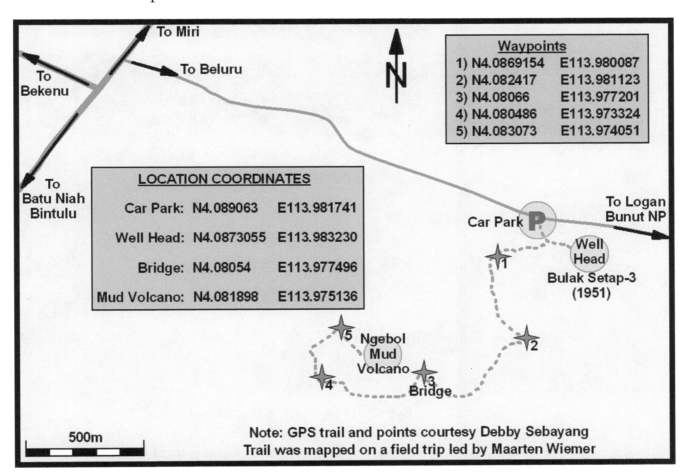

Figure 13.3.1a: Location map of the Ngebol mud volcanoes and Bulak Setap-3

3.13.2 Logistics and Safety

This visit is suitable for a half-day excursion. It is imperative to keep a safe distance from the active craters; in particular, large, low-lying and active craters should not be approached, for the mud is very fluid. There can be a false sense of security while walking on the hardened mud! Beware: it can be only a thin crust that will not withstand the weight of a person. In particular, children should be kept under close watch.

3.13.3 Outcrop Highlights

- Active mud volcanoes
- Overpressured Setap Shales
- Bulak Setap-3 gas and brine well

3.13.4 Geological Description

The mud volcanoes are located within the oil palm plantations (Figure 3.13.4a), on the road to Bintulu, near Bekenu Junction. Local people call the place "Ngebol", meaning "seepage". The larger area is called "Setap", a name that has been used to characterize the mudstone formation in the area.

Figure 3.13.4a: Overview of the Ngebol mud volcanoes

Mud volcanoes consist of generally low-relief surface mud mounds with a crater-like appearance (Figure 3.13.4b), produced by vertically migrating, cold fluidized mudstones and gases; they are not related to volcanic activity. The popping sound of the exploding gas bubbles is a characteristic feature of active mud volcanoes. Some water-filled low lying areas probably correspond to "calderas", i.e. depressions created by the explosion of mud volcanoes when larger volumes of gas expand at the surface (Figure 3.13.4c). Birds and wild animals frequent Ngebol in search of water; in the past, local people used to hunt close to the mud volcanoes.

Mud volcanoes are tiny pressure valves at the surface of the earth (Tingay et al., 2009). They are often located above the crest of geological structures (anticlines or diapirs) or zones of structural disturbance, where deep-seated, overpressured soft mudstones are associated with gas. The pressurized gas and water-rich mudstones make their way to the surface through areas of least resistance. During their rise, the gases mix with the muds and create a slurry. Close to the surface, the gas becomes more buoyant, rising more rapidly and erupting at the surface in the form of bubbles within the liquefied mud (Figure 3.13.4d). The constant upward flow of the slurry and subsequent hardening of the muds to the surface fashions the typical shape of the mud volcanoes. These features are transient, being constantly re-shaped by new mudflows. When they grow too high above the ground, mud volcanoes are often abandoned, the vents migrate to areas of less resistance, away from the mound. The viscosity of the mud determines the shape of the mound; in Ngebol, the water-rich muds create low and broad craters, reaching a maximum of one meter in height. The origin of the gases is mostly related to hydrocarbon generation in the subsurface, as evidenced from the frequent oil films seen around the mud volcanoes (Figure 3.13.4e). The gases consist usually of methane, which can be lit up by a flame.

Figure 3.13.4b: Mud volcanoes at Ngebol. Note erupting gas in the large water-filled crater in the back

Figure 3.13.4c: Emerged (forefront) and submerged (back) mud volcanoes. Note erupting gas in the large water-filled crater in the back

Figure 3.13.4d: Small gas bubble within dense mud

Figure 3.13.4e: Surface oil seeps at Ngebol

Mud volcanoes are common in many areas of Northern Sarawak, Brunei and Sabah where the Setap Shale outcrops. In the Jerudong Limbang area, shallow coreholes have identified various layers of liquid mud of 30-140 ft thicknesses, coinciding with a fault zone. In the Klias Peninsula, the higher subsurface pressures and the generally higher viscosity of the mudstones have resulted in a more violent type of mud volcanism. There, the manifestations can include huge explosions often lasting many days; large gas flares have also been reported. Such large volumes of rocks are expelled during these cycles of activity that whole islands have been created in this way (for example, Pulau Tiga, Sabah).

Figure 3.13.4f: Bulak Setap-3 well head

The route leading to Bulak Setap-3 well head has been described above. This well was drilled, cased, plugged back, cemented and abandoned as a non-commercial gas discovery in the early 1950s. Today, the cement plug inside the casing strings has cracked slightly and bubbling gas-rich brine flows out of the wellhead (Figure 3.13.4f). As the water is fairly clean, the flow proceeds probably from a deep aquifer in porous sandstone, and has no contribution from the Setap Shale proper. The deep aquifer is over-pressured, such that the water flows naturally upward. The gas is probably methane, generated inside the formation.

3.14 Liang Formation in Brunei

3.14.1 Overview

The predominant sand, clay and lignitic sediments of the Liang Formation were deposited during the Pliocene as part of the Baram Delta system. Their deposition is confined within embayments or protected coastal features overlying the troughs of the Badas and Berakas syncline in Brunei Darussalam, bounded by active north-south trending anticlines, not dissimilar to the present day Brunei Bay. This rugged coast probably restricted the influence of wave action and amplified tidal wave effects that may have created stacked thick sand accumulations, as is the case of the Tunggulian/Labi areas.

Sandstones of the Liang Formation have long been recognized as valuable materials for construction projects in Brunei (Sandal, 1996). Indeed, Liang Formation sandstones on the western flank of Lumut Hills were used as early as the 1950s for the construction of Ringer's Dyke in Seria (Harper, 1975). Sand excavation has shifted from the western flank of the Lumut Hills in the 1970s to the Tunggulian area (eastern flank of the Lumut Hills) in the 1990s. Total reserves of sandstone material at Tunggulian were estimated to be at 29.8 million cubic meters (Halim, Quazi Abdul, 1996) with production totaling about 2.3 million metric tons in 1996.

Outcrops of the Liang Formation can be best studied in two areas; the Tunggulian/Labi quarries in the Belait district, and the coastal areas of Berakas in the Brunei/Muara District (Figure 3.14.1a). The majority of the Tunggulian/Labi outcrops are in active quarries; fresh exposures show clearly the depositional systems and stratigraphic stacking patterns of the Liang Formation. Because of the fresh and unconsolidated nature of these outcrops there is a risk of landslides from unstable cliff faces and slopes; extra care should be taken when visiting the site. These active quarries are the least accessible of all the Liang Formation outcrops in Brunei Darussalam, but they are the closest to Miri. The outcrops of the Liang Formation in the Berakas coastal area are easily accessible and offer beautiful views of sandy beaches against the backdrop of the South China Sea.

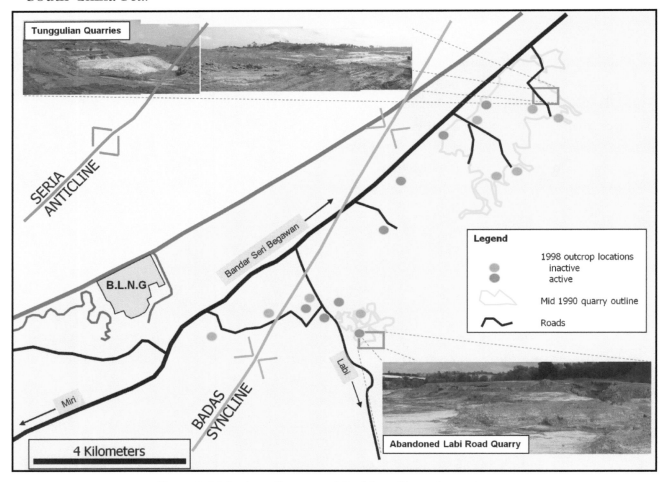

Figure 3.14.1a: Location map of the Liang Formation outcrops

To date, only a few comprehensive geological studies have been done on the Liang Formation. In the early 1990s, a broad geological study involving field mapping and shallow coring was made in order to delineate the distribution of sands of the Liang Formation in the Tunggulian area, prior to large-scale sand excavation projects (Geocon, 1992). A more recent study by the author has focused more on the stratigraphy and depositional models for the Liang Formation (Ibrahim, 1998).

Most of the current outcrops are constantly deteriorating due to fast weathering; some disappear entirely due to active excavation, while new ones appear elsewhere as a result of new digging. A guide to outcrop locations and generic descriptions of the associated stratigraphy and geology are presented in the following paragraphs.

3.14.2 Kampong Lalit Road Outcrop

The access road crosses the Village of Kampong Lalit and can be entered from the main road connecting Miri to Bandar Seri Begawan or via the Labi Road. Outcrops are located in a growing residential development area; exposures are accessible wherever a piece of land is excavated, particularly in the hilly areas. Here the sediments are interpreted to be the youngest part of the Liang Formation, and may be 2-3 million years old. Along this road, sand deposits rich in cemented fossils and shell fragments can be found. In some areas, the erosive base of these sands can be seen cutting into the more typical interbedded sand, shale and lignitic rich sequence of the lower Liang Formation.

3.14.3 Labi Road Outcrop

Within 4 miles from the junction with the main Miri to Bandar Seri Begawan, along Labi Road, the Liang Formation is exposed in various quarries, some abandoned and some active. The best outcrop location is an abandoned quarry in the vicinity of the JKR (Jabatan Kerja Raya) workshop, where a shale-rich sequence can be seen interbedded with thin sand beds, normally in the form of channel pods (Figure 3.14.3a). This sequence is also rich in lignites, resins, fossil rootlets and small tree stumps. Good examples of bioturbation and low energy depositional features such as flaser beds can also be observed here.

Thick sand deposits can also be found in outcrops, although their abundance both areally and stratigraphically was not observed to be as high as in the Tunggulian quarry areas. In the late 1990s some of the sands were freshly excavated, and the author observed similar sedimentary features within these sands as in those deposited in the Tunggulian area (tidal bundles, fining upward sequences, large vertical burrows with cemented lining).

Labi Road Sand Quarries

Type Lithology

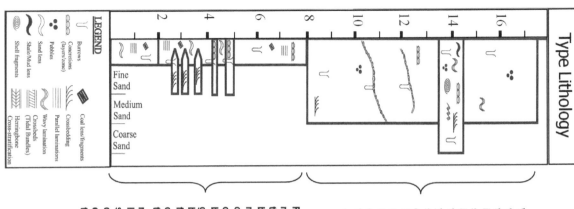

LEGEND

Burrows	Coal lens/fragments
Concretions (layers/zone)	Crossbedding
Pebbles	Parallel laminations
Sand lens	Wavy lamination
Shale/Mud lens	Crossbeds (Tidal Bundles)
Shell fragments	Herringbone Cross-stratification

Fine Sand

Medium Sand

Coarse Sand

Where this section is exposed, it is highly weathered. Sedimentary features are hard to distinguish, and sometimes can only be identified by the presence of cemented surfaces. These cemented surfaces indicates wavy and parallel laminations, the occasional trough crossbedding and if you're lucky, ripple marks. Abundant well sorted coarse grained and pebbly deposits, various types of concreted burrows and shell fragments are typically present, suggesting a high energy shallow marine depositional environment (beach or shoreface deposits). Such exposures that have been investigated were along Jalan Labi and Jalan Kampong Lalit.

Predominantly laminated shale rich sequence, interbedded with thin fine to medium grained sand bodies (H). The laminated shale rich beds are rich in lignitic clasts and bioturbation, while occasionally exhibiting flaser bedding (I). The sand bodies are generally crossbedded and bioturbated, with finer lignitic fragments sometimes deposited on the foresets (J). In general these sands do not extend very much laterally and display lens or channel like profiles. The best exposure of this sequence can be found in an abandoned quarry along Jalan Labi, where one may encounter debatable fossilised remains of tree stumps and roots (K & L).

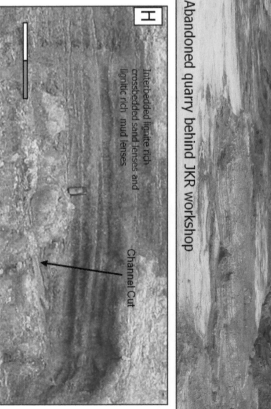

Abandoned quarry behind JKR workshop

H — Interbedded lignite rich crossbedded sand lenses and lignitic rich mud lenses — Channel Cut

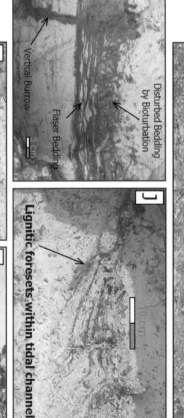

I — Vertical Burrow — Disturbed Bedding by Bioturbation — Flaser Bedding — 5 cm

J — Lignitic foresets within tidal channel

K — Roots — Fossilised Tree Stump

L — Fossilized tree stumps — 10 km

Figure 3.14.3a: Stratigraphic log and outcrop pictures of the Labi Road

3.14.4 Tunggulian Sand Quarries

The quarries here are easily observable from the air but quite hidden from the view onland as they are situated behind a series of low hills spanning the main road that connects Miri to Bandar Seri Begawan, reached via Tunggulian. Being under active exploitation, fresh outcrops are continuously exposed in the quarry (Figure 3.14.4a). Prior permission is required from the operating companies to enter and explore the quarries. While visiting the quarries one should be aware of the ongoing activities, and stay away from active operations. Practice utmost care and remain vigilant of your surroundings.

Large areas have been excavated and continue to provide fresh exposures, and we could make only sporadic short visits to study the outcrops. Here, the stratigraphic sequence of the Liang Formation consists of alternating thick highly bioturbated clean sand beds with black lignite, rich laminated shale beds and grey colored homogenous shale beds.

The sand beds vary in thickness reaching up to 14 m. Crossbedding is a common sedimentary feature of the sands, with some outcrops exhibiting clear symmetrical herringbone crossbeds. Burrows are abundant within these sands, particularly vertical burrows which indicate rapid deposition in a high-energy depositional environment. Channel profiles are sometimes recognizable; they normally have erosional bases cutting into more shale rich sequences at the base. These sand bodies seem to be elongated in a general east-west direction; paleocurrent flow directions measured in sedimentary features on a few outcrops confirm this.

In the Tunggulian area, three dominant shale lithofacies are observed in the Liang Formation; (1) interbedded lignite-rich shale and sand facies, (2) black laminated lignite-rich shale facies and (3) relatively homogenous, grey clean shale facies. The first facies is similar to that encountered in the Labi outcrops. The second facies is conspicuous by the presence of lignitic fragments that exhibit plant imprints on their layer's surfaces. The third facies is easily recognizable by the grey color of the shales and the presence of thin rootlets in an otherwise homogenous bed. All of these facies are highly bioturbated. These facies normally have transitional contacts with each other and the grey shales typically underlie the other facies. Each of these shale facies has thicknesses that vary from 30 cm up to 3 m.

The Liang Formation in the Tunggulian/Labi area has been interpreted as having been deposited as barrier islands with associated lagoons, tidal flats and marshes (Figure 3.14.4b). This explains the linear distribution of the clean, well-sorted sand deposits, as well as the contemporaneous presence of mud rich tidal influenced deposits related with the sands. Alternatively, these deposits could have been deposited within a protected coastal feature, such as an embayment or estuary, where wave action is reduced and tidal wave influence is relatively amplified instead. This could not only explain the distribution and presence of both the sands and associated shale facies but also their cyclical stacking pattern and prevailing paleocurrent flow directions. This alternative depositional model suggests active structuration and subsidence along the coast to continually create the protected coastal environment and space for the sediments to be deposited. The Liang Formation has been recorded to have a total thickness of 600 m (Sandal, 1996).

Tunggulian Sand Quarries

Type Lithology | **Type Stratigraphy**

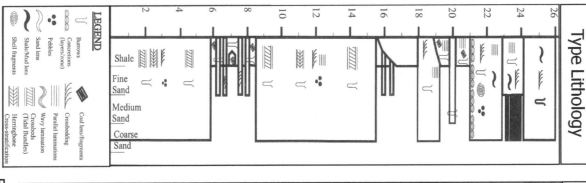

LEGEND

Burrows	Sand lens	Coal lens/fragments
Concretions (layers/zone)	Shale/Mud lens	Crossbedding
Pebbles	Wavy lamination	Parallel laminations
Shale fragments	Crossbeds (Tidal Bundles)	Herringbone Cross-stratification

Shale — Fine Sand — Medium Sand — Coarse Sand

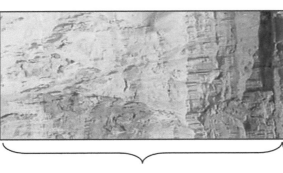

Sand Body. Large-scale bi-directional crossbeds, clean, predominantly medium to coarse grained, highly burrowed. Thin concreted basal unit. Thickness varies from 2-14m

Lignite-rich, predominantly shaley, with occasional interbedding with fine-grained sand laminated sequences. Contains woody fragments (A), plant imprints, resins, small burrows. Thickness varies from ca 0.3-3m

Pale grey relative softer mud-rich sequence (B) bed full of rootlets (C) (needle-like lignitic fragments/imprints). Typically transitional contact with lignite rich sequence while base can be cemented. Thickness varies from ca. 0.3-2m thick

Sand Body. Large-scale bi-directional crossbeds, clean, predominantly medium to coarse grained, highly burrowed. Thin concreted basal unit (D). Thickness varies from 2-14m

Some of these thick sand bodies exhibit prominent Herringbone crossbedding and extensive vertical burrows (tide dominated deposits; E & F) whilst others exhibit prominent trough crossbedding with prominent bioturbation, shell fragments and pebbly lag deposits (G). Evidence of channelisation present within these sand bodies. Some have erosional bases which are cemented present day.

Figure 3.14.4a: Stratigraphic log and outcrop pictures of the Tunggulian sand quarries

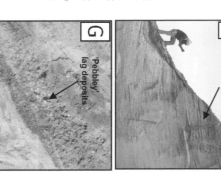 G 'Pebbley' lag deposits

 E 'Big' Burrow

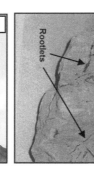 C Rootlets Rootlets

 A Plant Imprint

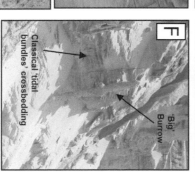 F Classical 'tidal bundles' crossbedding 'Big' Burrow

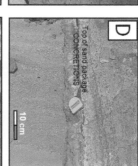 D Top of sand package CONCRETIONS 10 cm

 B

Depositional Models for the Tunggulian and Labi Liang Formation Deposits

There exists differing opinions as to the origin of the sand deposits in the Tunggulian and Labi area. Two alternative conventional depositional models are illustrated here, sharing a similar reliance on strong tidal influenced systems, both of which have analogues on the present day Northwest Borneo coast.

One model indicates the sands as being deposited in a barrier Island system, with abundant flood and tidal deltas as well as washover fans (M). Associated shale and lignite rich deposits seen in the outcrops are depicted as being deposited in the relatively low energy lagoonal, tidal flats and tidal marshes environments.

The other model suggests deposition of unusually sand rich tidal sand bars in a protected estuary (N), where tidal influence is amplified relative to wave influence. Structuration (uplift) in the vicinity of the estuary occasionally restrict the influx of water and sediments such that lagoonal or marshy depositional environment dominates, explaining the presence of cyclical mud and lignite rich sediments. Both models have supporting evidence in the outcrops, but the dominant east-west depositional direction (O) and widespread lateral North-South distribution of the sands give further credence towards the latter depositional model in the author's mind.

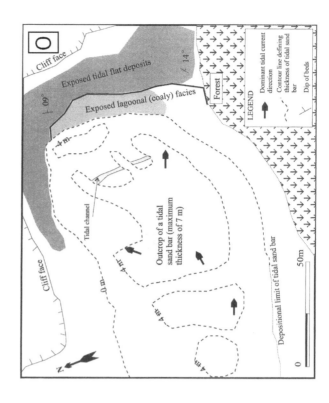

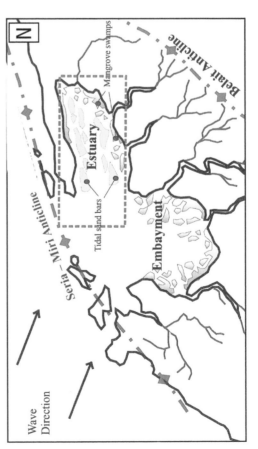

Figure 3.14.4b: Liang Formation depositional models for the Tunggulian and Labi deposits

3.14.5 Berakas Forest Reserve Beach

The Berakas Forest Reserve is located just off the main highway connecting Miri and Muara town. The area is popular with hikers due to the abundance of forest trails, and with picnickers for the beautiful sandy beach on the South China Sea. It also has miles of cliffs that expose the Liang Formation for the interested eye. These cliffs are easily accessible except during very high tides.

In the Berakas area, the Liang Formation can be distinctly divided into two lithofacies (Figure 3.14.5a). An older facies of interbedded sands and shales accumulated within sequences of tidal and distributary channels and associated overbank deposits that generally flow northwards. The second (younger) facies is comprised of interbedded sands, shales, conglomerates and entrained large lignitic fragments (including tree trunks). Large mature conglomerate can be found as isolated pods or beds or as a single rock engulfed within interbedded shales and sands matrix. This lithofacies corresponds to sedimentation in a relatively higher energy braided river system. These deposits are can be observed within a gully that cuts across the cliff faces on the western part of the Forest Reserve, requiring a short walk along the beach to reach it.

Similar active structuration is believed to have created an embayment in the Berakas area, allowing the deposition of tidally influenced distributary channels, protected from direct wave action of the South China Sea, probably by the actively growing Jerudong anticline in the west. Sometime during the Pliocene, further structural activity inland probably changed the drainage system and avulsed a higher energy river system through the Berakas area, depositing conglomeratic units near the present day coastal area. Similar conglomeratic deposits can be found further inland at exposures near the capital and in Temburong, providing a sediment source and supporting the depositional model presented above.

Liang Formation Deposits in the Berakas Coastal Area

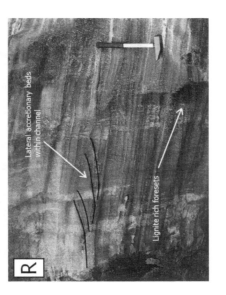

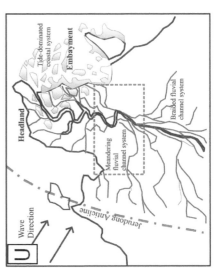

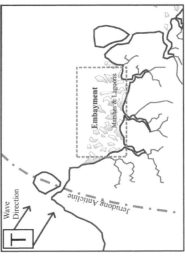

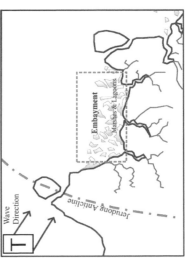

The Liang Formation can also be observed on miles of cliff exposures along the beautiful sandy beaches of the protected Berakas National Forest Reserve.

The deposits here are mainly interbedded sand and shale sequences of tidal and distributary channel deposits (P, Q & R. On a section of the cliff outcrops, higher energy braided river deposits can be found, typified by conglomeratic beds and entrained large lignified tree trunks (S). This braided river system has an overall erosive base, cutting through the tidal and distributary system deposits.

The Liang Formation here are believed to have been deposited within an embayment, similar to the present day Brunei Bay, created by the subsiding Berakas Syncline which is bordered by N-S trending Jerudong and Labuan Anticline (T). The braided river system was probably deposited when structuration further in the hinterland avulsed and forced drainage of a higher energy river system via this embayment (U).

Figure 3.14.5a: Overview of Liang Formation outcrops along the Berakas coastal area

187

3.15 Niah National Park and Caves

For a visit to Niah National Park, please consult the Sarawak Forestry Department's website at:
http://www.forestry.sarawak.gov.my/forweb/np/np/niah.htm
Further detailed descriptions of the park, its flora and fauna can be found in Hazebroek and Abang Morshidi (2000).

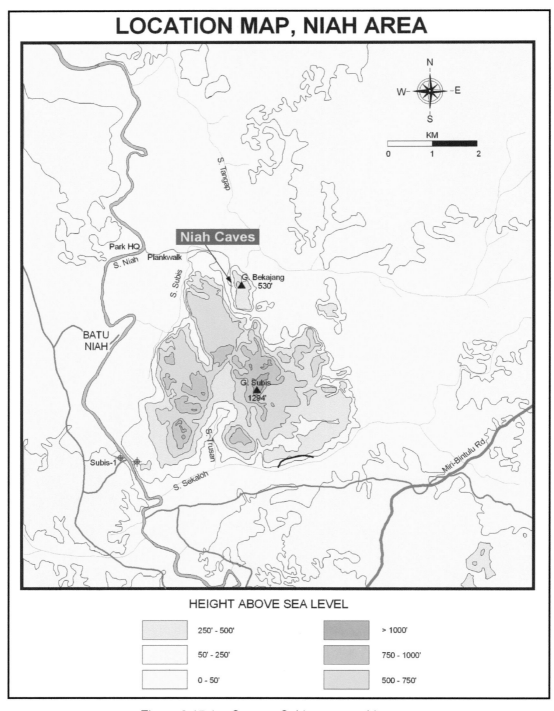

Figure 3.15.1a: Gunung Subis topographic map

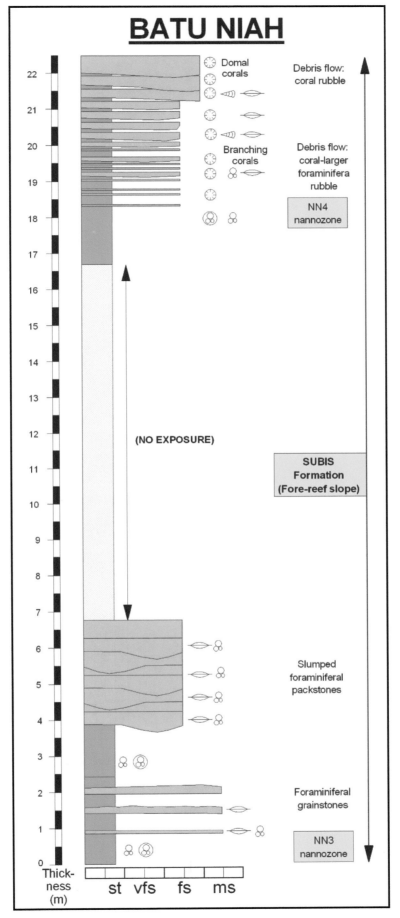

Figure 3.15.1b: Stratigraphic log of Subis Formation, Batu Niah

3.15.1 Overview

Rising abruptly out of the flat landscape (Figure 3.15.1a), the Niah Mountain, or the Subis Complex, provides a stunning view of how the seabed was shaped, some 20-15 million years ago! As you look at the mountain, imagine that the sealevel at the time was at the height of the highest peak; the base of that submarine mountain was seated deep under water. Driving towards Niah, the offshore marine sediments that were deposited around this island, the dark blue shales of the Setap Formation can be seen on road sides. The Niah Mountain itself is made up of limestones (the Subis Limestone) that built up from the seabed, as corals and other framework-building organisms were growing upward, keeping up with rising sea level. Corals can be seen in the active quarry at the entrance of Niah, where rocks are freshly excavated. Other outcrops are located along the road to Batu Niah (Figure 3.15.1b), where nannofossils indicate an age of 19-14.5 Ma (Lower to Middle Miocene, biozones NN3 and NN4). In the National Park, the limestones are superficially dissolved and coated with a veneer of microcrystalline carbonates due to the release of organic acids by the vegetation; the original components of the rock are therefore not visible at the surface.

The walkway from the Park Headquarters to the cave starts in a lowland tropical rainforest (mixed dipterocarp forest), but soon, the near vertical edge of the massive limestone is reached. This high cliff is an analogue to the Layang Layang drop off outside the reef of Labuan Island (located in the South China Sea, offshore Sabah): it

Figure 3.15.1c: Traders' Cave overhang

is the preserved reef wall of an island, isolated in the Early to Middle Miocene seas, some 20-15 million years ago! The Subis Complex is of mid-size, compared with the many similar structures buried offshore on the Central Luconia Platform, where large resources of oil and gas have been discovered. Micropaleontological data indicate that the Subis limestone is slightly older than the Luconia carbonates.

The Niah Caves are situated in a cliff on the northern part of the complex: they have been created by the slow dissolution of limestone by slightly acidic waters passing through the rock. The process of karstification, i.e. the formation of caves in limestones, is due primarily to the vertical filtration of water along joints and faults, down to a level where rocks are fully saturated with water (raising the water table), where caves will form. Due to recent uplift, this karst system is exhumed in Niah and the water table is now at the base of the complex, where the Niah River flows.

The Niah Cave complex is a typical perched, dry cave, or dead cave, meaning that no active stream runs through it and that the cave is in open atmospheric contact with the outside; such caves can only evolve by collapse. The first "cave" to be penetrated, the Trader's Cave, is actually a huge overhang. The curved shape of the overhang, particularly in its upper part, follows the path of an old meandering stream (Figure 3.15.1c). Similar curved shapes can be observed in the Deer Cave at Mulu, where a river is still actively flowing. At Niah, meandering galleries are further encountered, when crossing the Great Cave on the way to the Painted Cave. The outer edge of the Trader's Cave is lined by stalactitic tuffas and columns, which develop along a network

of hanging roots (Figure 3.15.1d). These soft, porous, heterogeneous and organic-rich deposits have a bulbous shape reminiscent of stalactites; however, their genesis is completely different from cave dripstones. Tuffa constructions form very rapidly in full daylight, contrarily to cave speleothems.

The dimensions of the Great Cave give an indication of the huge quantities of water and the length of time that were necessary to evacuate, by slow dissolution, the gigantic volume of rock now missing (Figure 3.15.1e). At the entrance, thick units of red claystones form a soft soil. These claystones are the insoluble residues left over from the dissolution of the limestones, that is to say, the impurities –the clay and iron-rich components- that constitute a minor fraction in the composition of the limestones. The archaeological site at the left entrance of the Great Cave is excavated in red soil and guano (Figure 3.15.1f); a 40,000 year old human skull was recovered there.

Of particular geological interest in the cave are the various roof openings, the sinkholes, which give an indication of how thin the roof of the cave is in places (Figure 3.15.1g). Associated with the sinkholes are the roof collapse blocks, which litter the floor of the cave: these are witnesses to the brief but drastic episodes when parts of the walls and the roof suddenly give way and cave in.

Figure 3.15.1d: Stalactitic tuffas and columns, Traders' Cave. Note the curved-shaped edge of the overhang

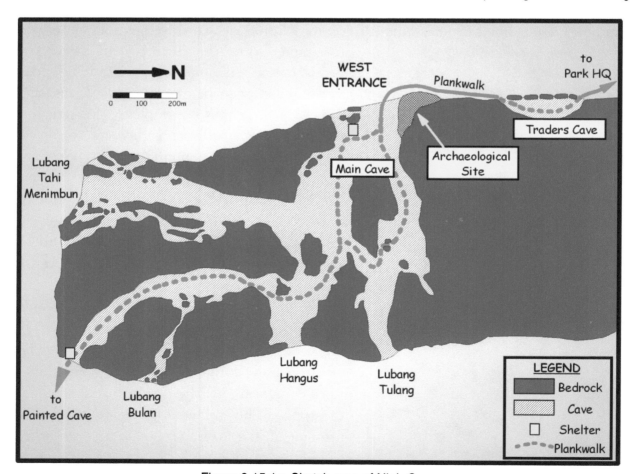

Figure 3.15.1e: Sketch map of Niah Cave

Figure 3.15.1f: Entrance to Niah's Great Cave. The archaeological site is on the right

Figure 3.15.1g: Sink hole, Niah's Great Cave   Figure 3.15.1h: Tuffa helictites, Painted Cave

Following the plankwalk and turning to the right at the bifurcation, down the steep staircase, one walks through an abandoned, underground meandering stream leading to a small perched valley. The Painted Cave is located at a short distance on the opposite flank of the valley. A rare tuffa helictite of large proportions can be observed in a fissure on the left hand cliff, shortly before reaching the Painted Cave (Figure 3.15.1h). Of geological interest here is the flat roof of the cave that follows the base of a homogeneous limestone bed (Figure 3.15.1i).

Returning on the plankwalk and following straight at the bifurcation, one climbs to the upper reaches of the cave system. At the top of the staircase, one passes at the base of a giant, hanging limestone column, some 15-20 m high (Figure 3.15.1j). There are scallop marks, indicating erosion by fast flowing waters, but the genesis of this suspended pillar is enigmatic; it is possibly linked to a system of joints. From this location, the path takes you back to the main entrance of the Great Cave.

Figure 3.15.1i: Horizontal bedding forms the flat roof of Pained Cave

Figure 3.15.1j: Hanging limestone column, Niah's Great Cave

3.16 Gunung Mulu National Park and Caves

For a visit to Gunung Mulu National Park, please consult the Sarawak Forestry Department's website at:
http://www.forestry.sarawak.gov.my/forweb/np/np/mulu.htm
For more technical details, consult the website of the Mulu Cave Project at:
http://www.mulucaves.org/
Further detailed descriptions of the speleological expeditions, the park, its flora and fauna can be found in Meredith et al. (1992) and Hazebroek and Abang Morshidi (2000).

3.16.1 Overview

Mulu is reached by plane from Miri; the journey offers opportunities to observe the large meanders of the Baram River, including abandoned arms of the river (oxbow lakes), in the flat alluvial plain (Figure 3.16.1a). The mountain landscape that suddenly arises from the plain creates a dramatic contrast; here we meet kilometer-high, near-vertical walls forming the edges of giant limestone mountains (Figure 3.16.1b), huge sinkholes, deep and narrow valleys, and weather permitting, a sight of the pinnacles.

Figure 3.16.1a: Aerial photograph of meanders, Baram River. Note the abandoned meanders with stagnant water

The limestone formation at Mulu is called the Melinau Limestone; it is a much older formation than the Subis Limestone at Niah. The oldest limestones in Mulu are about twice older than those of Niah, dating back 40-35 Ma. Limestones have been accumulating here for some 20-15 million years, creating a huge carbonate complex, over 2 km thick in parts (Wannier, 2009). The formation of caves is a very recent feature, as it started less than 2 million years ago, and is still going on today (Farrant et al., 1995). Gunung Mulu, the high mountain silhouette in the background, consists of even older rocks, the Mulu Formation, bridging the Late Cretaceous with the Early Tertiary (80-40 Ma). Here the rocks consist of shales, sandstones and conglomerates with abundant quartz deposited in a deep marine basin. This basin was squeezed and inverted into a submarine mountain at

Figure 3.16.1b: Aerial photograph of inaccessible steep limestone cliffs, Gunung Mulu National Park

45-40 Ma, and it is on this new, slowly subsiding marine shelf that the Melinau Limestone was deposited.

Both the Mulu Formation and the overlying Melinau Formation are tilted towards the north; running water that drains Gunung Mulu infiltrates the Melinau limestones, creating a system of subterraneous rivers and a giant karst system consisting of sometimes oversized caves (including the Sarawak Chamber, the largest cave in the world), huge single cave passages (Deer Cave has the largest single passage known on Earth, being 170 m wide and 120 m high), and a long network of interconnected galleries (including the Clearwater System, the eighth longest of its kind in the world with over 175 km of mapped galleries).

The phenomenal size of the karst system in Mulu is due to a combination of several factors: Very high rainfall, warm temperatures, and thick tropical vegetation cover result in high concentration of organic acids and CO_2. The cave system is located below the drainage area of Gunung Mulu, where aggressive water flows through the ground in large quantities. Today, the drainage pattern originating from Gunung Mulu is fairly mature, with the main streams bypassing the limestone massif, but in the past most of the runoff water penetrated the limestone complex.

Each of the 4 show caves at Mulu illustrates a different and complementary aspect of the limestone cave systems, and the plankwalk leading to the caves provides additional insight into the workings of the karst system.

Figure 3.16.1c: Inactive resurgence at the base of a limestone cliff; pathway to Deer Cave

The plankwalk from the Park Headquarters to the Deer and Lang caves passes, at first, through a lowland riverine forest, with minor limestone outcrops. Immediately after the long bridge over Sungai Melinau Paku, the pathway follows the edge of the massive limestone, and various resurgences can be seen along the way, some active and others abandoned. Resurgences are springs at the low end of a karst system: water reappears on the surface after a tortuous journey through the limestone. One such resurgence, now abandoned, can be observed shortly before arriving at the bat observatory; it is a truly spectacular sight, with a narrow chasm cut in the rocks at the base of the cliff (Figure 3.16.1c), and a long, rectilinear dry stream bed (Figure 3.16.1d) with large imbricated boulders in its prolongation. Here, water must have resurged as a jet under great pressure and volume, to cause the imbrications of such coarse rock fragments and the straight streambed. The pebbles and boulders that pave the dry streambed all come from Gunung Mulu: they have been transported through the limestone by powerful river systems!

Deer Cave is a cathedral in the rocks: its sheer dimensions and the magnificence of the setting are breathtaking (Figure 3.16.1e). Large volumes of water over long periods of time have created this monumental void. Today, a trickle of water continues flowing through the cave and follows the meanders cut on the side of the cave by the more powerful ancient rivers. As one enters Deer Cave, huge roof collapse blocks almost impede the passage (Figure 3.16.1f); inside the cave, indications of the northward dipping limestone beds can be seen on the right wall (Figure 3.16.1g). There is a general absence of speleothems in Deer Cave: the cave is a well-ventilated single solution tube, unsuitable for the development of such mineral deposits. The opening

opening towards the Garden of Eden, at the rear end of the cave, shows the entrance of a small stream originating from the slope of Gunung Mulu (Figure 3.16.1h). The huge rounded boulders and large size pebbles that line up the bed of this stream are stranded remnants of a much more powerful river that once engulfed this cave.

Figure 3.16.1d: Straight stream boulder and pebble bed in front of inactive resurgence (3.16.1c)

Figure 3.16.1e: Entrance to Deer Cave

Figure 3.16.1f: Huge collapsed roof blocks at Deer Cave entrance

Figure 3.16.1g: Inside Deer Cave, the largest cave passage on Earth

Figure 3.16.1h: The Garden of Eden, at the rear end of Deer Cave

Cyclic climate variations from glacial to interglacial and back to glacial have marked the last 2 million years; in the tropics, climate variations fluctuated from abundant rains to relatively dry climates. Much of the karst features in Mulu was formed during periods of extreme rainfall.

Lang Cave is the smallest of the show caves in Mulu, but it is rich in speleothems of all sorts; as the roof of this cave is seldom over a few meters high, it is ideal to make detailed observations of these mineral constructions. Speleothems are created as water saturated with calcium carbonate slowly emerges in the cave from the pores of the rock; release of pressure and evaporation result in the precipitation of calcite within the cave. This process can lead to various shapes being formed: dripstones, straws or stalactites, hanging from the roof of the cave (Figure 3.16.1i); stalagmite towers growing upward on the floor (Figure 3.16.1j); columns, when stalagmites and stalactites coalesce (Figure 3.16.1k); flowstones include hanging draperies (Figure 3.16.1l) and calcitic curtains (Figure 3.16.1m) and sculptured mounds on the walls (Figure 3.16.1n) or on the cave floor; helictites, developing laterally in a twisted fashion (Figure 3.16.1o).

The visit to Wind and Clearwater caves is done by boat going up Sungai Melinau. At the entrance of Wind Cave, stalactitic tuffas can be observed, as they develop along a network of hanging roots.

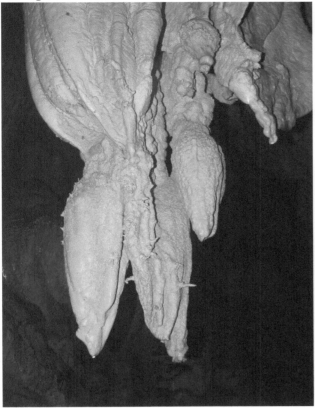

Figure 3.16.1i: Bulbous stalactites with secondary growth of small helictites, Lang Cave

Figure 3.16.1j: Stalagmite with complex growth structure, Lang Cave

Figure 3.16.1k: Column, formed by the connection of stalagmite and stalactite, Lang Cave

Figure 3.16.1l: Flowstones and draperies, Lang Cave

Figure 3.16.1m: Folded roof draperies, Lang Cave

Figure 3.16.1n: Flowstones, Lang Cave

Figure 3.16.1o: Bulbous stalactites with overgrowth of smaller helictites, Lang
Cave. Note the alignment of the stalagmites along a fracture zone in the roof

Figure 3.16.1p: Entrance gallery, Wind Cave

Figure 3.16.1q: Massive flowstones lining the entrance walls of King's Room, Wind Cave

The entrance into Wind Cave follows an old riverbed gallery, a tubular passage formed along a north-dipping limestone bed (Figure 3.16.1p). After a narrow labyrinthic passage through scalloped limestone walls (typical for high water flow conditions), one gets access to a sinkhole. From there, the path leads down into King's Room, where larger flowstone formations can be seen (Figure 3.16.1q). Much of the floor of that chamber consists of blocks fallen from the roof; remarkably, vertically-standing stalagmites have been able to grow on strongly tilted blocks (Figures 3.16.1r)! King's Room is

Room is rich in speleothems of all sorts, particularly stalagmites with complex growth figures (Figure 3.16.1s) and large columns (Figure 3.16.1t).

Figure 3.16.1r: Vertically growing stalagmites on highly tilted roof-collapse block, King's Room, Wind Cave

Figure 3.16.1s: Variations in dripping rates led to the formation of complex stalagmites, some of which exhibit higher rates of growth at their tops, as compared to their base; King's Room, Wind Cave

Figure 3.16.1t: Column formed by the junction of an
oversized and high stalagmite, and a minute stalactite,
King's Room, Wind Cave

Figure 3.16.1u: Resurgence at valley-level, Clearwater Cave

Picture 3.16.1v: Clearwater Cave entrance along dipping limestone beds (top left of picture)

Figure 3.16.1w: *Monophyllaea*-covered tuffa stalactites, Clearwater cave

Figure 3.16.1x: Large subterraneous river in its gallery, Clearwater Cave

Clearwater Cave, a short distance away, has an entrance, high above the valley floor and a deep system of multiple galleries, the deepest one being the site of an active river, which resurges below the cave entrance at the level of Sungai Melinau (Figure 3.16.1u). The wall at the mouth of the cave (Figure 3.16.1v) is lined up with meter-long tuffa stalactites, covered with *Monophyllaea* single-leafed plants (Figure 3.16.1w). Inside the cave, the path leading to the subterraneous river level zigzags through collapsed blocks from the roof. The carved gallery through which the river flows has impressive dimensions: 30m width and 30m height (Figure 3.16.1x)! The path follows the river upward for a little while, before reaching a sinkhole, where it stops.

Located at walking distance from the Royal Mulu resort, the "hot springs" are worth a visit. This small resurgence bubbles with gas, including hydrogen sulfide. The origin of this gas is possibly volcanic or linked to the cracking of organic matter (kerogen) in the subsurface sediments. It is possible that sulphuric acids are present along the buried flowing stream, where they may dissolve the limestone formations and create/enlarge a cave system.

Butterflies, including the beautiful "Rajah Brooke" can be frequently observed at this location.

Chapter 4

Collecting Fossils

4.1 About Fossils

4.1.1 What are Fossils?

Most animals do not die of old age, but due to predation, disease, and natural disasters (e.g. storms that deposit them onto the beach or bury them deeply within sediments, and so on). Moreover, dead animals decay and are eaten very quickly, unless they are buried immediately under an impermeable sedimentary layer. Even then, except for insects trapped in amber, it is not possible for soft parts of animals to be preserved. For all these reasons, living beings have rare chances to become fossils, and generally only their harder parts (mostly skeletons) can be preserved and left behind. The vast majority of these harder parts are crushed, dissolved, broken down by waves, etc. Only a very small number of animal parts are thus preserved, and that only because they are protected by the overlying sediments.

In fine-grained sediments (for example mud) chemically reducing conditions cause the skeletons to loose their original colors very quickly and instead become bluish or grayish. This can be observed by a very simple experiment: Put a few dead mollusks (with the animals in their shells topped up with water) in an airtight vessel. After only a few days the colors of the shells will have disappeared – the shells may have a color between dirty white and dark grey. Watch out if you open the bag – there will be an awful smell from the hydrogen sulphide produced during decomposition. Coarser sediments generally contain more oxygen and the original colors of the skeletons may thus be preserved or they may also obtain yellowish color.

These are the first steps in the process of fossilization. Subsequently, sediments containing the animal remains are compacted, water in the pores of sediments is expelled, and chemical reactions cause the sediment particles to become cemented. These chemical reactions also cause many of the shells to fracture (e.g. Plate 4.1.1a, Figures 1, 2 and 4), as the sediments turn into a hard rock. Fluids generally continue to flow through the rocks, and in time, may dissolve the shells, leaving only the sedimentary rock with the original shell as a mold; this is common place in the sandy layers of outcrops around Miri (e.g. Plate 4.1.1a, Figure 3).

4.1.2 Collecting Fossils

Rock outcrops in the Miri area often crumble because of rainfall; therefore, fossils can generally be collected without damaging outcrops. It is important, however, to always respect the outcrop and minimize your impact, such that outcrops can survive longer for future visitors.

Many of the Sarawak fossil shells have been partially dissolved, and are, therefore, fragile. They fall apart quickly when the rocks are eroded by rain or coastal waters, or when cleaned too briskly by the collector. For handling such fragile fossils a soft brush should be utilized, using no or very little water. The fossils enclosed in harder concretions are more robust (Figure 4.1.2a). Fragile specimens can be strengthened by applying a mixture of soluble glue and acetone – but good ventilation is required. Ensure that your fossil samples have proper labels on them, recording the locality, date, and stratigraphic level from which they were collected. Also write the name of the collector and the scientific name of the species, if you can identify it.

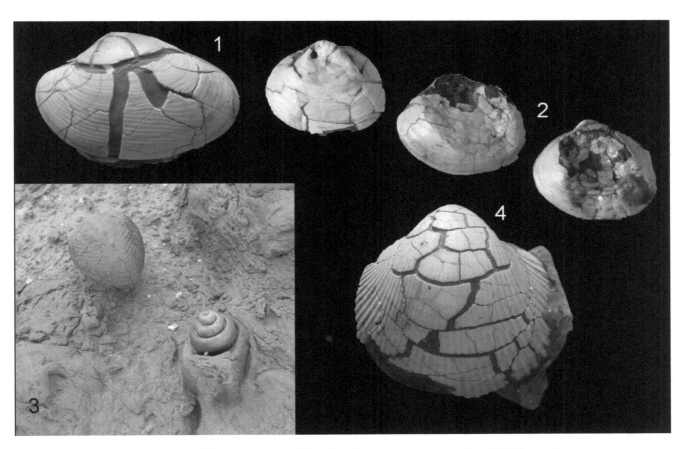

Plate 4.1.1a: Different states of fossilization: examples from the Miri Formation
Figure 1: *Paphia neglecta* (Martin, 1919) – specimen showing cracked valve – Beraya-Bekenu road; Figure 2: *Cycladicama oblonga* (Hanley, 1844) – three specimens showing crushing of the left valve (seen in numerous specimens of this species) - Beraya-Tusan sea cliffs km 8.3 (3.60); Figure 3: *Vepricardium (Hemicardium) njalindungense* (Martin, 1922) (top left) and *Cymatium* spec. (bottom right) – moulds - Beraya-Tusan sea cliffs km 9.5 (2.35); Figure 4: *Vepricardium (Hemicardium) njalindungense* (Martin, 1922) – showing cracked valve - - Beraya-Tusan sea cliffs km 10.7 (1.2).
(See chapter 4.6.4 for references to distance)

Figure 4.1.2a: Detailed view of fossilized seabed surface with bivalves and fossil debris, Beraya-Tusan seacliffs

4.1.3 Microfossils and the Dating of Rocks

Large fossils (macro-fossils, which can be seen with unaided eye) are sometimes abundant in marine sediments, but most sedimentary beds are devoid of macrofossils. In contrast, a high diversity and abundance of microfossils can be retrieved from marine sedimentary rocks and observed under a microscope. They include minute crystalline lattices secreted by algae, the so-called nannoplankton (Figure 4.1.3a); tiny organic particles such as cysts of dinoflagellates, spores, sporomorphs and pollen of land-plants blown out to sea by the wind (Figure 4.1.3b); and a group of marine unicellular animals with an external shell - called foraminifera (Figure 4.1.3c). Some foraminifera evolved to gigantic shapes, with shells several centimeters long. The limestone formations at Niah and Mulu contain such macro-foraminifera, often in great abundance (Figure 4.1.3d).

Animals and plants have changed their forms as they evolved through geologic times, generally becoming more complex in their shapes and functions. Paleontologists have painstakingly reconstructed many aspects of how and when these evolutionary changes in the shapes and sizes of organisms occurred in the past. So, by identifying key fossils, it is now possible to know the geological age of fossils and, by extension, the age of the rocks in which they reside.

Microfossil studies have been pursued by the oil industry for many decades because they enable the geologist to correlate sedimentary formations in a given petroleum field or across sedimentary basins. This technique was introduced in the 1930s and immediately applied to the geological study of the Miri oil field. A succession of different microfossil markers (biozones of foraminifera) was identified and used to separate the various productive sandstone formations that make up the Miri field (Figure 2.3.3a).

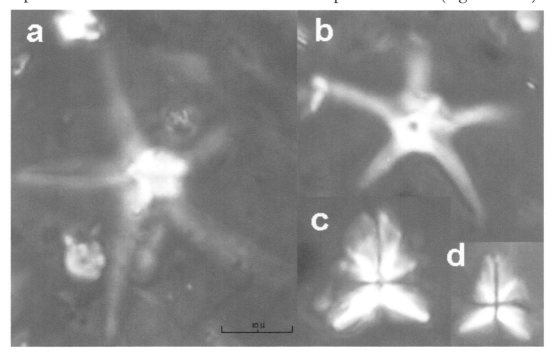

Figure 4.1.3a: Nannoplankton remains consist of calcium carbonate plates formed by single-cell algae. Specimens from the Upper Miocene Liang Formation in Tutong (Brunei), courtesy of Musa Musbah. a,b) *Discoaster quinqueramus*; c) *Sphaenolithus abies*; d) *Sphaenolithus neoabies*.

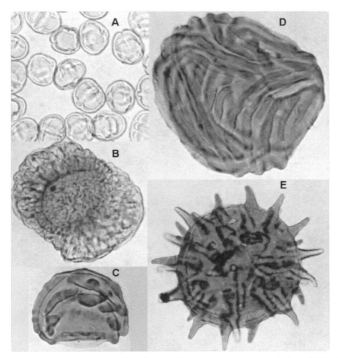

Figure 4.1.3b: Pollen and spores, examples from Brunei, courtesy of Jan Wilschut. a) *Rhizophora*, mangrove swamp vegetation; b) example of pollen with airsacks; c) bean-shaped spore *Stenochlaena laurifolia*; d) Fern spore *Ceratopteris thalictroides;* e) Characteristic pollen type derived from *Hibiscus tiliaceus*, a common tree in the beach forests and along the coast of Borneo.

Figure 4.1.3c: Foraminifera, examples from Sarawak.
a) arenaceous-type benthonic foraminifer; b)
porcellaneous-type benthonic foraminifer; c) hyaline-
type benthonic foraminifer; d) planktonic foraminifer

Figure 4.1.3d: Larger benthonic foraminifer from the Lower Miocene, Mulu

4.1.4 Trace Fossils

Marine sediments deposited on the sea floor often loose their sedimentary structures, as animals burrow and move through the sediments for a safe shelter and for search of food. Sedimentary layers with a disturbed or homogenized structure due to the passage of animals are said to be bioturbated. This is particularly frequent in shallow marine environments, where the largest diversity and abundance of animals are found, thanks to a large supply of food, mainly plankton.

Figure 4.1.4a: Trace fossil *Ophiomorpha nodosa* has characteristic walls, lined with layers of sand pellets, forming a 3-dimensional network of linked galleries within the sediment; Tanjung Lobang

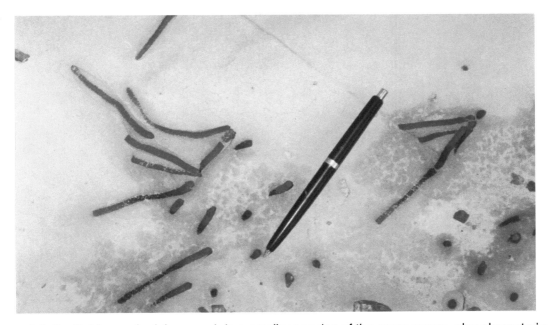

Figure 4.1.4b: *Ophiomorpha labuanensis* is a smaller species of the same genus, also characterized by pellet-lined walls; Pujut Petronas station

Figure 4.1.4c: *Teichichnus* trace fossils consist of vertically-stacked, elongated burrows;
Jalan Lopeng

Figure 4.1.4d: *Gyrolithes* trace fossils consist of an helicoidal tunnel, burrowed vertically
within the soft sediment; Jalan Padang Kerbau Outcrop 3

In such "bioturbated intervals," the animals themselves are not preserved as fossils, but traces of their movements and passage remain. Crabs, in particular, construct a 3-dimensional network of galleries within sediments, for which they strengthen the walls with secretions, as in the case of *Ophiomorpha* (Figures 4.1.4a,b). Other animals create simple vertical tunnels, allowing them to reach the seabed for feeding, and to retract within the sediment for security. Trace fossils are important indicators of conditions that prevailed at the seabed interface; often, the occurrence of particular types of trace fossils can help reconstruct the depositional environment and water depth at which the sediments were deposited. The shallow marine sediments around Miri contain a rich and diverse assemblage of trace fossils, for which we illustrate a few characteristic examples in Figures 4.1.4a-e.

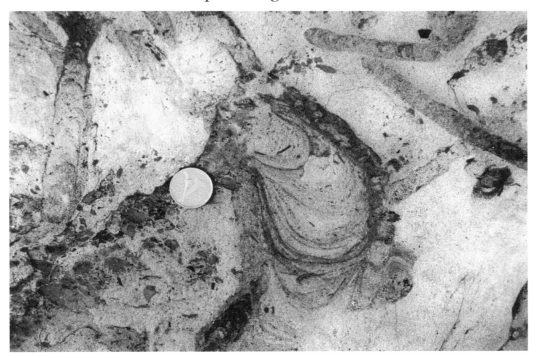

Figure 4.1.4e: *Rhizocorallium*, a U-shaped trace fossil with internal concave laminae (spreite); Tusan Angus

4.2 Fossil Research around Miri

The rich Miocene and Pliocene assemblages of marine fossils from the seacliffs south of Miri were first noted by geologists working for Bataafsche Petroleum Maatschappij in the 1930s. They sent fossil collections for age determination to Professor Dr. K. Martin at Leiden, a task that was continued later by Dr. C. Beets at what is now called Netherlands Centre for Biodiversity Naturalis in Leiden. The results of these studies were used for petroleum exploration in Sarawak and Brunei. Regrettably, only hand written reports of these studies were produced and descriptions of only few species (mollusks) were eventually published (probably also due to lack of precise stratigraphic information). Four species were described by Beets: from Sarawak, *Brechites (Brechites) venustulus* (in 1942; see Figure 16 on Plate 4.5.4e) and from Brunei, *Cardilia bruneiana* (1941), Acila *bruneiana* (1942) and *Ringicula serianensis* (1987). Nuttall (1961) published a list of fossil species from outcrops in Brunei. A few characteristic species from the Liang Formation are described in Sandal et al. (1996) and some species from the Miri and Seria Formations in Brunei are shown in Raven (2008).

The occurrence of marine fossils in the Miri area was first published in the early 1960s (Liechti et al., 1960 and Wilford, 1961), based on materials collected by the Shell geologists and by the Geological Survey Department during the 1950s. Parts of these early collections, now held at the Natural History Museum in London, form the basis of the first comprehensive review and publication on the fossil crabs from NW Borneo by Morris & Collins (1991). Collins described a total of 36 species of fossil crabs collected from 17 localities across NW Borneo, only two of which are in Sarawak. Recent efforts by Charlie Lee (1996-2009) have vastly expanded the collection of fossil crabs from Sarawak, with more than 5,000 specimens collected to date. A small but representative sample of this new collection has been deposited in the Natural History Museum in London, and the results of their analyses were published by Collins et al. (2003). At least 31 species of fossil crabs, from 24 genera, have been recorded from Miri. Two of these crabs were named after their discovery locations in Miri: *Phylira trusanensis* (after Tanjung Tusan, also named "Trusan" on older maps) and *Mursia bekenuensis* (after the town of Bekenu).

Charlie Lee and Han Raven have also collected fossil mollusks from the Miri outcrops. Description of the already known and newly discovered species are given in Raven, 2002 , 2008; Vermeij & Raven, 2009) or in papers in preparation.

The crustacean and mollusk faunas from the Liang and Miri Formations indicate shallower water-depth conditions than those of the Sibuti Formation. These fossils document an overall shallowing upward trend consistent with the regional geology. Demonstration of this trend by fossil crabs include *Mursia bekenuensis* (50-1300 m water depth) in the Sibuti Formation (Plate 4.7.4b, Figure 1); *Galena obscura* (< 100 m water depth and *Nucia platyspinosa* (< 50 m water depth) in the Miri Formation (Plate 4.5.4c, Figure 10; Plate 4.5.4b Figure 1); *Macropthalmus (Mareotis) woodwardi* and *Scylla serrata*, both found in tidal mudflats and mangrove swamps in the Liang Formation (Plate 4.4.4a Figures 6 & 8). Fossil mollusks indicate the same trend, e.g. *Haustellum mindanaoensis* (7-280 m water depth; Plate 4.7.4c, Figure 2) in the Sibuti Formation; *Distorsio decipiens* (50-100 m; Plate 4.5.4d, Figure 16) in the Miri Formation; and *Gafrarium pectinatum* (0-20 m; Plate 4.4.4b, Figure. 5) in the Liang Formation.

4.3 Fossil Localities around Miri

The main fossil localities are located in the Beraya-Bekenu-Sibuti area, south of Miri City. Here, rock formations comprise a thick sequence of sandstones and mudstones deposited in a deltaic to marine environments 10-20 million years ago. These sediments form part of the Setap Shale, Sibuti Formation, Lambir Formation and the Miri Formation (Figure 2.3a).

Marine fossils are common in the Setap Shale, Sibuti and Miri Formations. The majority of fossil-rich intervals in these formations are in mudstone beds. This is not surprising because mudstones are generally deposited in low-energy environments and are thus suitable for preservation of organic remains as fossils, while in higher-energy sandstone beds, shells easily break or dissolve. The fossil-rich beds in these mudstones have a distinctive bluish-grey color, and often contain numerous brown, nodular siderite concretions. The fossils occur in two distinct forms, either as loose individuals or fragments on the weathered surface of the outcrops or in concretions and concretion beds. In some localities fossils are well preserved, sometimes with the original shell intact, and very rarely with the delicate internal structures also preserved.

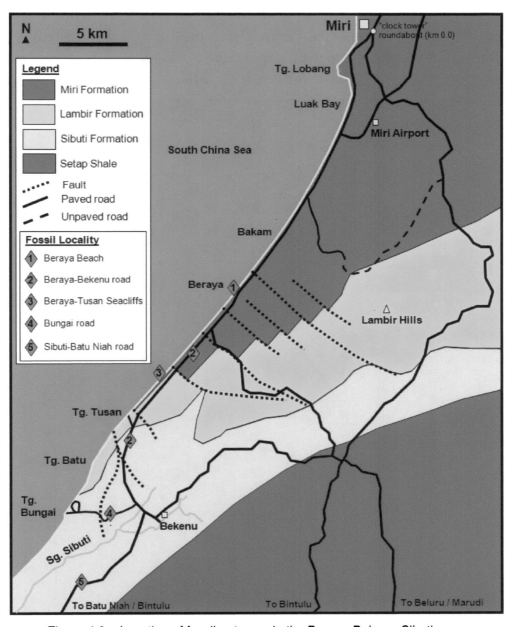

Figure 4.3a: Location of fossil outcrops in the Beraya-Bekenu-Sibuti area

In the Beraya-Bekenu-Sibuti area, the fossil-bearing outcrops are concentrated in five localities (Figure 4.3a), along or in the vicinity of the Miri-Bintulu coastal road: (1) Beraya Beach, (2) Beraya-Bekenu road, (3) Beraya-Tusan seacliffs, (4) Bungai road, and (5) Sibuti-Batu Niah road. At each of these locations (except for the Beraya Beach), there are several fossil-bearing outcrops, either in roadcuts, excavations or seacliffs.

4.4 Beraya Beach

4.4.1 Access and Location

The best way to explore Beraya Beach is on foot. From Miri town centre (Clock Tower roundabout, km 0.0), the drive to the Beraya Beach turnoff (no sign posted) is about 24 km (Figure 4.4.1a). The beach, less than 100 m from the main road, is accessible via a dirt ramp at the end of the road. Explore the beach on either side of Sungai Beraya for fossils.

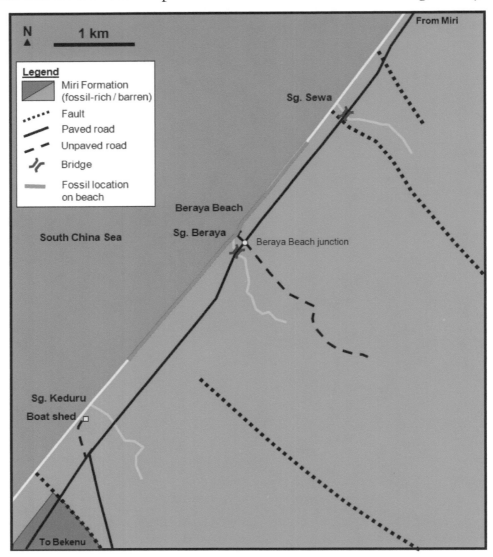

Figure 4.4.1a: Beraya Beach location map

4.4.2 Logistics and Safety

The beach is best explored at low tide (1.0 m or less). Watch out for sharp objects that may accumulate at the high tide mark. As there is no shade on the beach, ensure sufficient sun protection and carry plenty of drinking water.

4.4.3 Outcrop Highlights

•   Fossils on deserted beaches (Figure 4.4.3a), particularly a unique record of fossil crabs: *Macrophthalmus sp.* (Plate 4.4.4a, Figure 6) are reported from this locality

Figure 4.4.3a: Fossil crabs on Beraya Beach

4.4.4 Geological Description

Beraya Beach, located about 24 km southwest of Miri, is known for the occurrence of fossil crabs (Plate 4.4.4a) and other fossils on the beach, first noted in 1996 by Charlie Lee. The fossils are generally found among beach debris on the high tide mark along a two-kilometer stretch of the beach centered on Sungai Beraya. The presence of borings by marine organisms and the rounded edges of some fossils indicate an extended submergence in the sea and subsequent re-working and transportation by current and wave-action before deposition on the beach. A few "Beraya-type" fossil crabs have been found as far southwest as the beaches around Sungai Uban, about seven kilometers away, suggesting transportation by long shore currents.

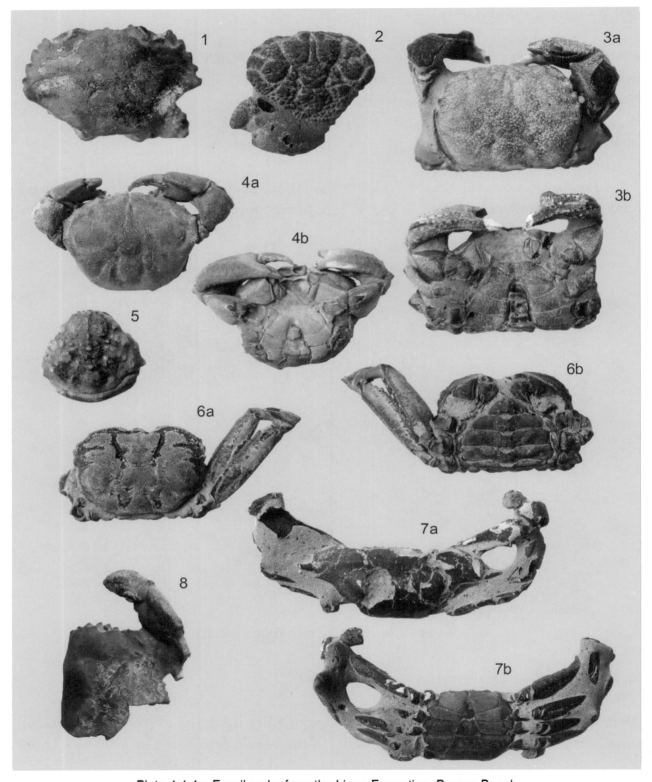

Plate 4.4.4a: Fossil crabs from the Liang Formation, Beraya Beach
Figure 1: *Charybdis anulata* (Fabricius, 1798), dorsal view; Figure 2: *Cyclodius ungulatus* (H. Milne Edwards, 1834), dorsal view; Figure 3: *Galena litoralis* (Collins, Lee & Noad, 2003), a-b, dorsal and ventral views; Figure 4: *Galena obscura* (H. Milne Edwards, 1873), a-b, dorsal and ventral views; Figure 5: *Pisoides bidentatus* (H. Milne Edwards, 1834), dorsal view; Figure 6: *Macrophthalmus (Mareotis) wilfordi* (Morris & Collins, 1991), a-b, dorsal and ventral views; Figure 7: *Podophthalmus vigil* (Fabricius, 1798), a-b, dorsal and ventral views; Figure 8: *Scylla serrata* (Forsskal, 1755), dorsal view with pincer.

The rock formation from which the fossils are derived has yet to be identified. The area surrounding Beraya Beach is made up of sandstones and mudstones of the Miri Formation. However, no fossils have been found in these nearby outcrops. It is likely that the fossil-bearing deposits are exposed in the subtidal area. Based on the presence of *Macrophthalmus* crabs (Plate 4.4.4a, Figure 6), which are probably Pliocene-Pleistocene in age, these deposits are equivalent of the Liang Formation from Brunei.

The most notable and abundant fossil found at Beraya Beach is the crab *Macrophthalmus (Mareotis) wilfordi*, a fossil found nowhere else around Miri. Fossil *Macrophthalmus* crabs are reported from many localities in the Indo-Pacific region (Sawata, 1991; Schweitzer, et. al., 2002), all from the Pliocene to Quaternary deposits. The closest examples to Miri are specimens from marine alluvium in Muara, Brunei (Wilford, 1961), Labuan and Kuala Padas in Sabah (Idris, 1989).

Other fossils found at Beraya Beach include bivalves, gastropods (Plate 4.4.4b), sea urchins, sand dollars, fish vertebra and eight other species of fossil crabs. Both the crab and mollusk faunas indicate a shallow marine to estuarine, tidal flat to mangrove swamp environment of deposition. More than 3,000 fossil crab remains have been collected from this site since 1996. The frequency of fossil finds on this beach has been decreasing in the last few years, likely indicating a depleting stock.

Plate 4.4.4b: Bivalves, gastropods and crabs from the Liang Formation, Beraya Beach
Bivalves: Figure 1: *Ostreidae* – genus and species still to be determined; Figure 2: *Atrina (Atrina) vexillum* (Born, 1778), fragment; Figure 3: *Circe (Circe) scripta* (Linnaeus, 1758); Figure 4: *Paphia neglecta* (Martin, 1919); Figure 5: *Gafrarium pectinatum* (Linnaeus, 1758).
Gastropods: Figure 6: *Murex altispira* (Ponder & Vokes, 1988); Figure 7: *Vexillum (Costellaria) subdivisum* (Gmelin, 1791) - the shell is largely dissolved; Figure 8: *Nassarius (Niotha) crenoliratus* (A.Adams, 1852) – various specimens in a concretion formed around a crab burrow.
Crab: Figure 9: *Macrophthalmus (Mareotis) wilfordi* (Morris & Collins, 1991), dorsal view

4.5 Beraya-Bekenu Road

4.5.1 Access and Location

From the Miri town centre (Clock Tower roundabout, km 0.0), the drive to the Beraya Batu Satu junction (sign posted) is about 27 km (Figure 4.3a). The start of the fossil-rich exposures is about 500 m southwest of the junction, on both sides of the road (km 27.5 to km 42.6; Figure 4.5.1a). Individual fossil outcrops are indicated in Figure 4.5.1b.

4.5.2 Logistics and Safety

Road safety is a particular concern at this location. The coastal road is narrow, often with soft and steep edges, and very limited space/location to park safely. To get to the outcrops some walking along the side of the road is necessary; so watch out for passing vehicles (often at high speed!), especially when crossing the road.

Figure 4.5.1a: View of fossil outcrop at km 28.3L, Beraya-Bekenu road

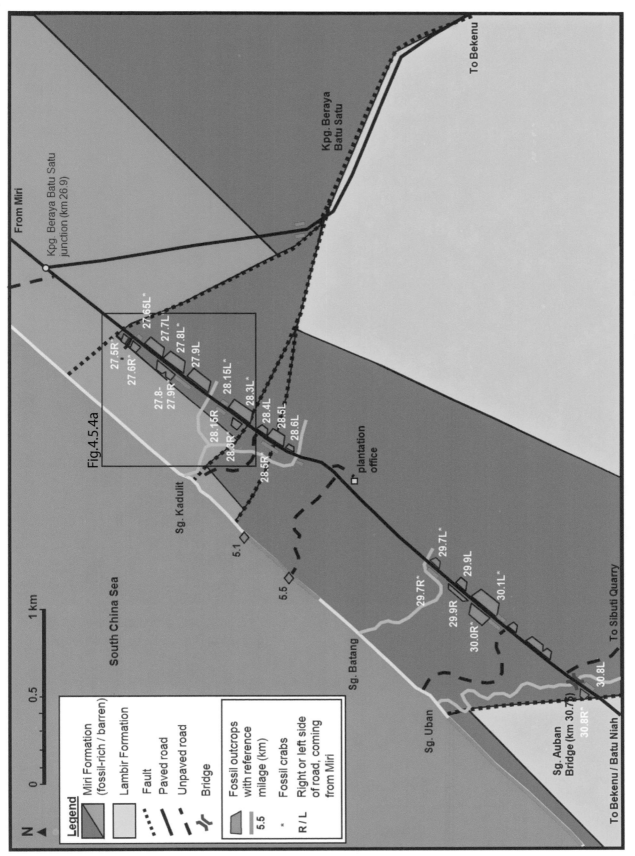

Figure 4.5.1b: Location map of fossil outcrops along Beraya-Bekenu road

4.5.3 Outcrop Highlights

This is the most accessible fossil site around Miri, with fossil-rich outcrops along the coastal road. Fossil shells and crabs, with the crab *Galena obscura* (Figure 4.5.3a) are commonly found between km 27.6 and 28.3 (Figure 4.5.4a). The highest fossil abundance and diversity is at kilometer 30.1 (Figure 4.5.3b). The mollusk and crab fauna indicates a clear shallowing of the environment from the Sibuti Formation upwards.

Figure 4.5.3a: Fossil crab *Galena obscura*; Beraya-Bekenu road km 27.7 left

Figure 4.5.3b: *Coquina* bed with bivalves and shell fragments; Beraya-Bekenu road km 30.1 left

4.5.4 Geological Description

The Beraya-Bekenu road cuts through rolling coastal ridges and hills, consisting of sandstones and mudstones of the Miri, Lambir and Sibuti Formations (Figure 4.3a). This stretch of the coastal road, constructed between 2002 and 2004, has exposed a thick sequence of fossil-rich mudstones (Miri Formation) very similar to that along the Beraya-Tusan sea cliffs. Fossil-rich beds can be seen on both sides of the road mainly between km 27.5 to 30.8 (Miri Formation) and km 37.7 to 42.6 (Sibuti Formation).

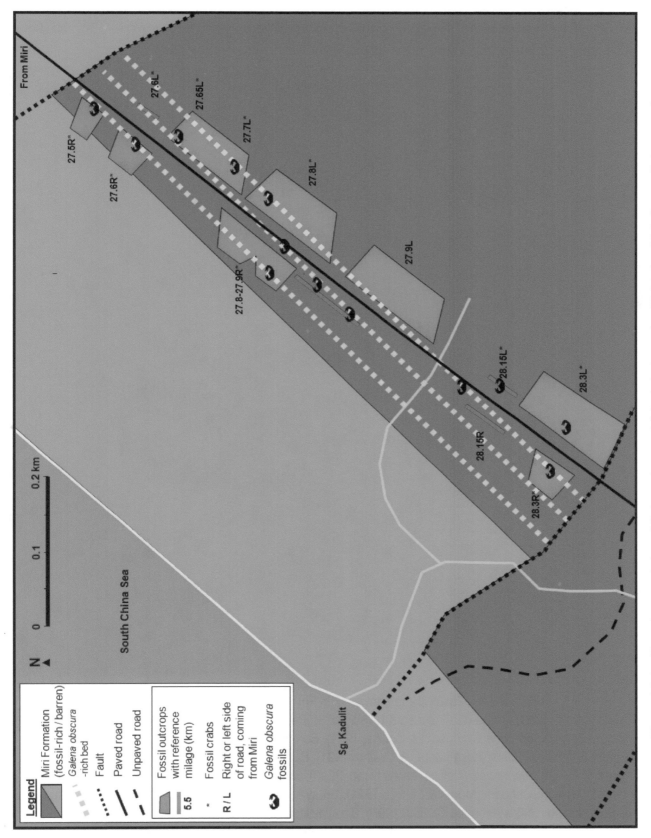

Figure 4.5.4a: Tentative correlation of Galena obscura-rich beds between km 27.6 to km 28.3, Beraya-Bekenu road

Mapping of the fossil outcrops indicates that the Beraya-Tusan area is controlled by a series of NW-SE trending transpressional faults (Figures 4.3a, 4.5.1b and 4.5.4a) repeating the same beds at multiple localities.

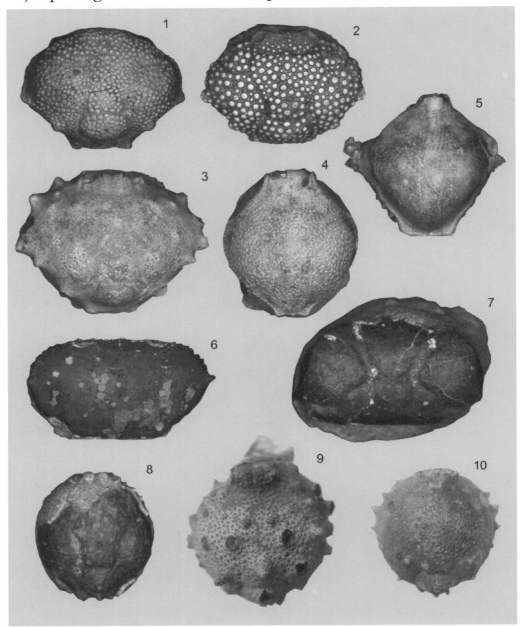

Plate 4.5.4b: Fossil crabs from the Miri Formation, Beraya-Bekenu road
Figure 1: *Nucia platyspinosa* (Collins, Lee & Noad, 2003), Beraya-Bekenu road km 29.7 right, dorsal view;
Figure 2: *Nucia borneoensis* (Morris & Collins, 1991), Beraya-Bekenu road km 29.7 right, dorsal view;
Figure 3: *Iphiculus sexspinosa* (Morris & Collins, 1991), Beraya-Bekenu road km 30.1 left, dorsal view;
Figure 4: *Philyra trusanensis* (Collins, Lee & Noad, 2003), Beraya-Bekenu road km 30.1 left, dorsal view;
Figure 5: *Myra subcarinata* (Morris & Collins, 1991), Beraya-Bekenu road, km 30.1 left dorsal view; Figure 6: *Portunus obvallatus* (Morris & Collins, 1991), Beraya-Bekenu road km 30.1 left, dorsal view; Figure 7: *Hexapus decapodus* (Morris & Collins, 1991), Beraya-Bekenu road km 30.1 left, dorsal view; Figure 8: *Nucilobus symmetricus* (Morris & Collins, 1991), Beraya-Bekenu road km 30.1 left, dorsal view; Figure 9: *Pariphiculus multituberculatus* (Collins, Lee & Noad, 2003), Beraya-Bekenu road km 29.9 right, dorsal view; Figure 10: *Pariphiculus papillosus* (Morris & Collins, 1991), Beraya-Bekenu road km 29.7 right, dorsal view.

Numerous species of crabs and mollusks have been collected from this locality. A few of the mollusks, corals, and crabs are illustrated on Plates 4.5.4b-e (together with fossils from the Beraya-Tusan seacliffs). Examination of the fauna is in progress.

Plate 4.5.4c: Fossil crabs from the Miri Formation, Beraya-Bekenu road

Figure 1: *Parthenope sublitoralis* (Morris & Collins, 1991), Beraya-Tusan seacliffs km 8.4 (Auban km 3.45), dorsal view; Figure 2: *Leucosia calcarata* (Collins, Lee & Noad, 2003), Beraya-Bekenu road km 29.9 right, dorsal view; Figure 3: *Drachiella guinotae* (Morris & Collins, 1991), Beraya-Bekenu road km 29.9 left, dorsal view; Figure 4: *Heikea tuberculata* (Morris & Collins, 1991), Beraya-Bekenu road km 30.1 left, dorsal view; Figure 5: *Portunus woodwardi* (Morris & Collins, 1991), Beraya-Bekenu road km 30.1 left, dorsal view; Figure 6: *Calappa sexaspinosa* (Morris & Collins, 1991), Beraya-Bekenu road km 28.5 left, dorsal view; Figure 7: *Charybdis annulata* (Fabricius, 1798), Beraya-Bekenu road km 30.1 left, dorsal view; Figure 8: Pincer of *Scylla serrata* (Forsskal, 1755), Beraya-Bekenu road km 30.1 left, dorsal view; Figure 9: Fish vertebra, fish tooth and shark's tooth, Beraya-Tusan seacliffs km 8.3 (Auban km 3.55); Figure 10: *Galena obscura* (H. Milne Edwards, 1873), Beraya-Bekenu road km 27.8 right, dorsal view; Figure 11: *Philyra granulosa* (Morris & Collins, 1991), Beraya-Tusan seacliffs km 8.3 (Auban, km 3.55), dorsal view.

Plate 4.5.4d: Gastropods from the Miri Formation, Beraya-Bekenu road and Beraya-Tusan cliffs
Figure 1: *Vycariella angsanana* (Martin, 1921) - Beraya-Tusan km 8.3 (3.60); Figure 2: *Haustator spec.* – Tanjung Batu; Figure 3: *Zaria javana* Martin, 1882 – Beraya-Tusan km 11.4 (0.4); Figure 4: *Turritella terebra spectrum* Reeve, 1849 – Beraya-Tusan, km 8.4 (3.45); Figure 5: *Vermetus spec.* - Beraya-Bekenu road km 28.5 left; Figure 6: *Xenophora (Stellaria) chinensis* (Philippi, 1841) – these animals agglutinate other shells and materials onto their shell for camouflage - Beraya-Tusan km 9.5 (2.35); Figure 7: *Natica (Natica) vitellus* (Linnaeus, 1758) – Beraya-Tusan, km 8.3 (3.60); Figure 8: *Natica (Natica) tigrina* (Röding, 1798) – Beraya-Tusan km 8.3 (3.60); Figure 9: *Natica rostalina Jenkins*, 1863 – with a drill hole made by a carnivorous snail - Beraya-Tusan km 8.3 (3.60); Figure 10: *Natica rostalina Jenkins*, 1863 – operculum - Beraya-Tusan km 8.3 (3.60); Figure 11: *Sinum (Sinum) eximium* (Reeve, 1864) – Beraya-Tusan km 8.3 (3.60); Figure 12: *Barycypraea murisimilis* (Martin, 1879) – Beraya-Tusan km 9.9 (2.0); Figure 13: *Varicospira javana* (Martin, 1899) – Beraya-Tusan km 8.3 (3.60); Figure 14: *Bursina nobilis* (Reeve, 1844) - Beraya-Tusan km 8.3 (3.60); Figure 15: *Phalium (Semicassis) bisulcatum* (Schubert & Wagner, 1829) – Beraya-Tusan km 8.3 (3.60); Figure 16: *Distorsio (Distorsio) decipiens* Reeve, 1844 – Beraya-Tusan km 8.3 (3.60); Figure 17: *Gyrineum lacunatum* (Mighels, 1845) – Beraya-Tusan km 8.3 (3.60); Figure 18: *Pterynotus (Pterynotus) alatus* (Röding, 1798) – Beraya-Tusan km 8.3 (3.60); Figure 19: *Murex brevispina macgillivrayi* (Dohrn, 1862) – Beraya-Tusan km 8.4 (3.45); Figure 20: *Xancus rembangensis* (Pannekoek, 1936) – Beraya-Tusan km 9.9 (2.0); Figure 21: *Phos (Phos) acuminatum* (Martin, 1879) - Beraya-Bekenu road km 28.5 left; Figure 22: *Peristernia bandongensis* (Martin, 1881) - Beraya-Bekenu road km 28.5 left; Figure 23: *Clavilithes verbeeki* (Martin, 1895) - Beraya-Tusan km 8.4 (3.45); Figure 24: *Vexillum (Costellaria) gembacanum* (Martin, 1884) - Beraya-Bekenu road km 28.5 left; Figure 25: *Agaronia (Anazola) gibbosa* (Born, 1778) – Beraya-Tusan km 8.3 (3.55); Figure 26: *Melongena gigas* (Martin, 1883) – Beraya-Tusan km 3.70 – fragment of the upper part of the whorls; Figure 27: *Cymbiolum multiplicatum* (Pannekoek, 1936) – Beraya-Tusan km 8.3 (3.60); Figure 28: *Architectonica perspectiva* (Linnaeus, 1758) – Beraya-Tusan km 8.3 (3.60)

Plate 4.5.4e: Bivalves and corals from the Miri Formation, Beraya-Bekenu road and Beraya-Tusan cliffs Figure 1: *Anadara (Anadara) scapha* (Linnaeus, 1758) - Beraya-Tusan km 9.5 (2.35); Figure 2: *Trisidos (Trisidos) torta* (Mörch, 1850) – Beraya-Tusan – no layer; Figure 3: *Cucullaea cf. pamotinensis* Martin, 1910 – Beraya-Tusan km 9.5 (2.35); Figure 4: *Amussiopecten singkirensis* (Martin, 1909) – Beraya-Tusan km 11.2-11.4 (0.4 to 0.6); Figure 5: *Spondylus sondeianus* Martin, 1909 - Beraya-Bekenu road km 30.1 left; Figure 6: *Hyotissa hyotis* (Linnaeus, 1758) – Beraya-Tusan km 11.4 (0.4) ; Figure 7: *Cycladicama oblonga* (Hanley, 1844) – Beraya-Tusan km 8.3 (3.60); Figure 8: *Vepricardium (Hemicardium) njalindungense* (Martin, 1922)- Beraya-Tusan km 10.7 (1.2); Figure 9: *Gari (Psammobia) preangerensis* (Martin, 1922) – Beraya-Tusan km 10.7 (1.2); Figure 10: *Solecurtus spec.* – Beraya-Tusan; no layer; Figure 11: *Leptomya (Leptomya) cochlearis* (Hinds, 1844) – Beraya-Tusan km 10.7 (1.1); Figure 12: *Clementia (Clementia) papyracea* (Gray, 1825) – Beraya-Tusan km 8.4 (3.45); Figure 13: *Dosinia (Dosinella) trailii* (A. Adams, 1855)– Beraya-Tusan km 8.2 (3.70); Figure 14: *Dosinia (Sinodia) insularum* Fischer-Piette & Delmas, 1967 - Beraya-Bekenu road km 27.8 left; Figure 15: *Paphia spec.* –Beraya– Tusan km8.4 (3.45); Figure 16: *Brechites (Brechites) venustulus* Beets, 1942 – Holotype – Tanjong Batu; Figure 17: *Gastrochaena (Gastrochaena) cymbium* (Spengler, 1783) – these animals generally live bored inside other shells: the arrows indicate the holes through which they extrude their siphons - Beraya-Tusan km 9.5 (2.35); Figure 18: *Corbula (Notocorbula) tjiguhanensis* (Martin, 1922) – Beraya-Tusan km 8.3 (3.60); Figure 19: *Cycloseris sp. (Fungiidae)* solitary coral with coral gall crab (?) cavity; Figure 20: *Pseudosiderstrea (Siderastreidae) coral.*

233

4.6 Beraya-Tusan Sea Cliffs (Auban)

4.6.1 Access and Location

The best way to explore the Beraya-Tusan Seacliffs is by driving along the beach during low tides (1.0 m or less). From the Miri town centre (Clock Tower roundabout, km 0.0), the drive to the Beraya Beach turnoff (no sign posted) is about 24 km (Figure 4.3a and 4.4.1a). The beach, less than 100 m from the main road, is accessible via a dirt ramp at the end of the road. Cross Sungai Beraya (km 0.0) and drive southwest for about five kilometer to reach the start of the start of the fossil-bearing outcrops (km 5.1 to km 11.5, Figures 4.5.1b and 4.6.1a). Individual fossil outcrops are indicated in Figure 4.6.1b. The fault contact at km 11.5 separates the fossil-bearing Miri Formation from the Lambir Formation. From here, drive another two and a half kilometer further south-west to reach Tanjung Tusan, a prominent rocky headland. An alternate access to this stretch of the coast is from Tanjung Tusan (road end at cliff top, Figure 4.6.1b). Park your car at the end of the road and walk or cycle to the fossil locations.

Figure 4.6.1a: Fossiliferous concretion bed at Beraya-Tusan Seacliffs

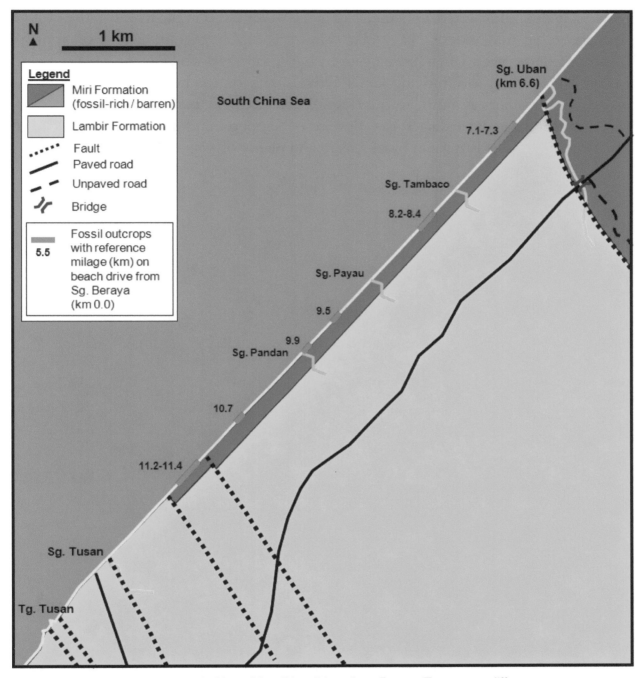

Figure 4.6.1b: Map of fossil localities along Beraya Tusan sea cliffs

4.6.2 Logistics and Safety

Beach drive is suitable for vehicles with high clearance and 4WD only. Ensure you have a full tank of gasoline, carry enough drinking water and other necessities, and check the tide level. The nearest gas station is in Miri or Bekenu, 20-24 km from Beraya Beach! Driving conditions on the beach is dependent on the weather, waves and tides. In good weather it is possible to drive this stretch of beach at tides of one meter or lower. There are five main river crossings on the beach (Figures 4.4.1a, 4.5.1b, 4.6.1b and 4.6.2a,b), with many other unmarked, ephemeral streams. Water levels in the river can rise rapidly with incoming tide and/or heavy rain. Always check, by wading across on foot, if unsure. Reduce speed when crossing river, but never stop while in the river to avoid getting stuck in the soft, moving sands. Drive slowly on the beach and watch out for wave/current scours and drift woods/logs. When stopping, park your vehicle far from the cliff face to avoid falling rocks. Beware of slippery mud and the sharp parts of burrows and fossils! Bring sun block cream and plenty of water.

Figure 4.6.2a: River crossing along Beraya-Tusan beach

Figure 4.6.2b: Crossing Sungai Keduru at high tide

Figure 4.6.3a: View of Beraya-Tusan Seacliffs (km9.5)

4.6.3 Outcrop Highlights

Here you will see spectacular seacliffs (Figures 4.6.3a), with small waterfalls and locally abundant vegetation, deserted beaches and plenty of fossils. Noteworthy fossils include large, palm size crabs *(Charybdis sp.)* at km 8.4 and km 9.9 (Figure 4.6.3b); cowries *Barycypraea murisimilis* (Plate 4.5.4d, Figure 12) at km 9.5 and km 9.9; *Zaria javana* (Plate 4.5.4d, Figure 3) in concretions at km 7.1, km 7.3 and km 11.2; large gastropods and bivalves at km 9.5 (Figure 4.6.3c; Plate 4.5.4d, Figure 6; Plate 4.5.4e, Figures 1, 3 and 17).

Figure 4.6.3b: Concretion with fossil crab *Charybdis annulata*; Beraya-Tusan Seacliffs, km9.9

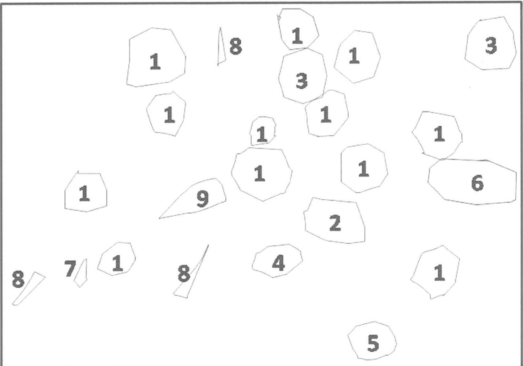

Figure 4.6.3c: Block of hardened sediment with molluscs, Miri Formation - Beraya-Tusan km 9.5 (2.35).
Bivalves. Figure 1: *Vepricardium (Hemicardium) njalindungense* (Martin, 1922); Figure 2: *Macoma (s.l.) spec.*; Figure 3: *Dosinia spec.*; Figure 4: *Paphia neglecta* (Martin, 1919); Figure 5: *Paphia spec.*; Figure 6: *Lutraria (Lutrophora) complanate* (Gmelin, 1791); Gastropods. Figure 7: *Vycariella angsanana* (Martin, 1921); Figure 8: *Haustator spec.*; Figure 9: *Conus spec.*

4.6.4 Geological Description

The Beraya-Tusan seacliffs (Figure 4.3a), part of a 15-kilometers stretch between Beraya and Bungai, are made up of sandstones and mudstones of the Lambir and Miri Formations. Originally laid down horizontally under the sea, these sediments have been folded and eroded, and they now form steeply dipping cliffs and ridges along the coast. A series of late transpressional faults dissected these rock formations, displacing them further inland towards the northeast, culminating at the Lambir Hills (465 m).

Along the Beraya-Tusan coast, the Miri Formation forms a series of low cliffs, dipping 30-50 degrees NW and extending into the South China Sea. The fossil-bearing intervals in the Miri Formation, here predominantly consisting of mudstones, are exposed discontinuously along an 8-kilometers stretch of coastline between Sungai Kadulit and Sungai Tusan (Figures 4.5.1b and 4.6.1b). Detailed mapping of the fossil outcrops indicate that the Beraya-Tusan area is controlled by a series of NW-SE trending transpressional faults (Figures 4.3a and 4.5.1b).

This fossil locality has also been referred to as "Auban Seacliffs" (Lesslar & Wannier, 1998; Raven, 2002; Vermeij & Raven, 2009) – using a kilometer reference from a marker point that has since disappeared. In this field guidebook distances are measured from the Clock Tower roundabout in Miri but the seacliff distances have been measured from Sungai Beraya on the beach; on the legends for the plates the original reference is indicated between parentheses (note that the old reference point was the southern side –near Tanjung Tusan- of the seacliffs and the new reference is to the northern side).

A number of specific beds can be differentiated based on mollusk fauna, which are likely correlatable over short distances. Specific beds recognized so far are:
- Bed with abundant *Amussiopecten singkirensis* (km 11.4; Plate 4.5.4e, Figure 4).
- Bed with *Zaria javana* (km 11.2; Plate 4.5.4d, Figure 3). The same species occurs in smaller numbers in the outcrops from 7.10-7.30 km which are several kilometers away, but stratigraphically only slightly younger.
- Bed with *Anadara pilula* (km 5.1; Figure 4.6.4a).

Work is in progress to correlate the outcrops of the Miri Formation in the Beraya-Tusan seacliffs and the Beraya-Bekenu road localities. The stratigraphic sequence appears to be (from youngest to oldest):
- The road section between km 27.5 and km 28.3
- The *Anadara pilula* layer outcropping in the seacliffs (Figure 4.6.4a) and along the Beraya-Bekenu road at km 28.4 L.
- The road section between km 29.7 and km 30.8
- The beds from km 7.10 to km 10.7 in the seacliffs (e.g. Figure 4.6.3a)
- The most basal layers are the *Amussiopecten singkirensis* and *Zaria javana* beds exposed in the seacliffs (km 11.2 to km 11.5); not exposed along the road section

A few of the mollusks and crabs are illustrated on Plates 4.5.4b-e (together with fossils from the Beraya-Bekenu road). Work on the analysis of the fauna is in progress.

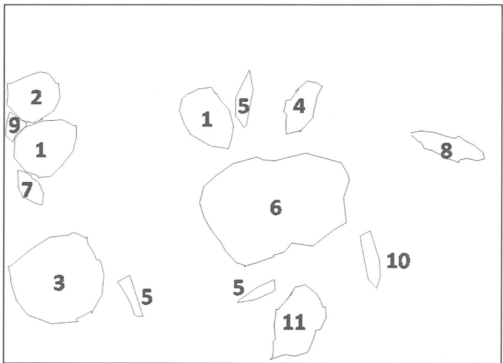

Figure 4.6.4a: Block of hardened sediment with molluscs, Miri Formation - Beraya-Tusan km 5.1 (6.65); same fauna as Beraya-Bekenu road km 28.4 left.

Bivalves. Figure 1: *Anadara (Potiarca) pilula* (Reeve, 1843); Figure 2: *Anomia spec.*; Figure 3: *Vepricardium (Hemicardium) njalindungense*; Figure 4: *Pitar tusanensis* Beets

Gastropods. Figure 5: *Haustator cingulifera* (G.B. Sowerby I, 1825); Figure 6: *Xenophora chinensis* (Philippi, 1841); Figure 7: *"Ziba" spec.*; Figure 8: *Vexillum (Costellaria) gembacanum* (Martin, 1884)

Others. Figure 9: bryozoan; Figure 10: crab leg; Figure 11: iron staining around crab pincer.

4.7 Bungai Road

4.7.1 Access and Location

From the Miri town centre (Clock Tower roundabout, km 0.0), the drive to the Bungai Road junction (sign posted for Bungai Beach on the right) is about 43 km along the coastal road (Figure 4.3a). From this junction (km 0.0), there are several fossil-rich outcrops on both sides of the road, between km 1.4 to km 2.5 (Figure 4.7.1a). The slope leading to the outcrop at km 2.5 (Right hand-side) is thickly overgrown. The road ends at Bungai Beach, about three and half kilometers further west.

4.7.2 Logistics and Safety

Approach to the outcrop at km 2.5R is overgrown and the site itself is secluded, being hidden from the road by thick vegetation.

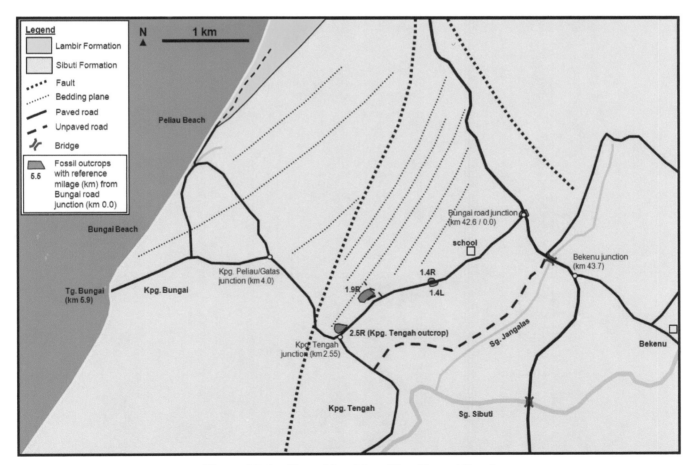

Figure 4.7.1a: Map of fossil localities, Bungai Road

4.7.3 Outcrop Highlights

This is an extremely rich outcrop, with fossil crabs *Mursia bekenuensis*, palm-sized sand dollars, small tubular and mushroom corals, and fish and shark's teeth.

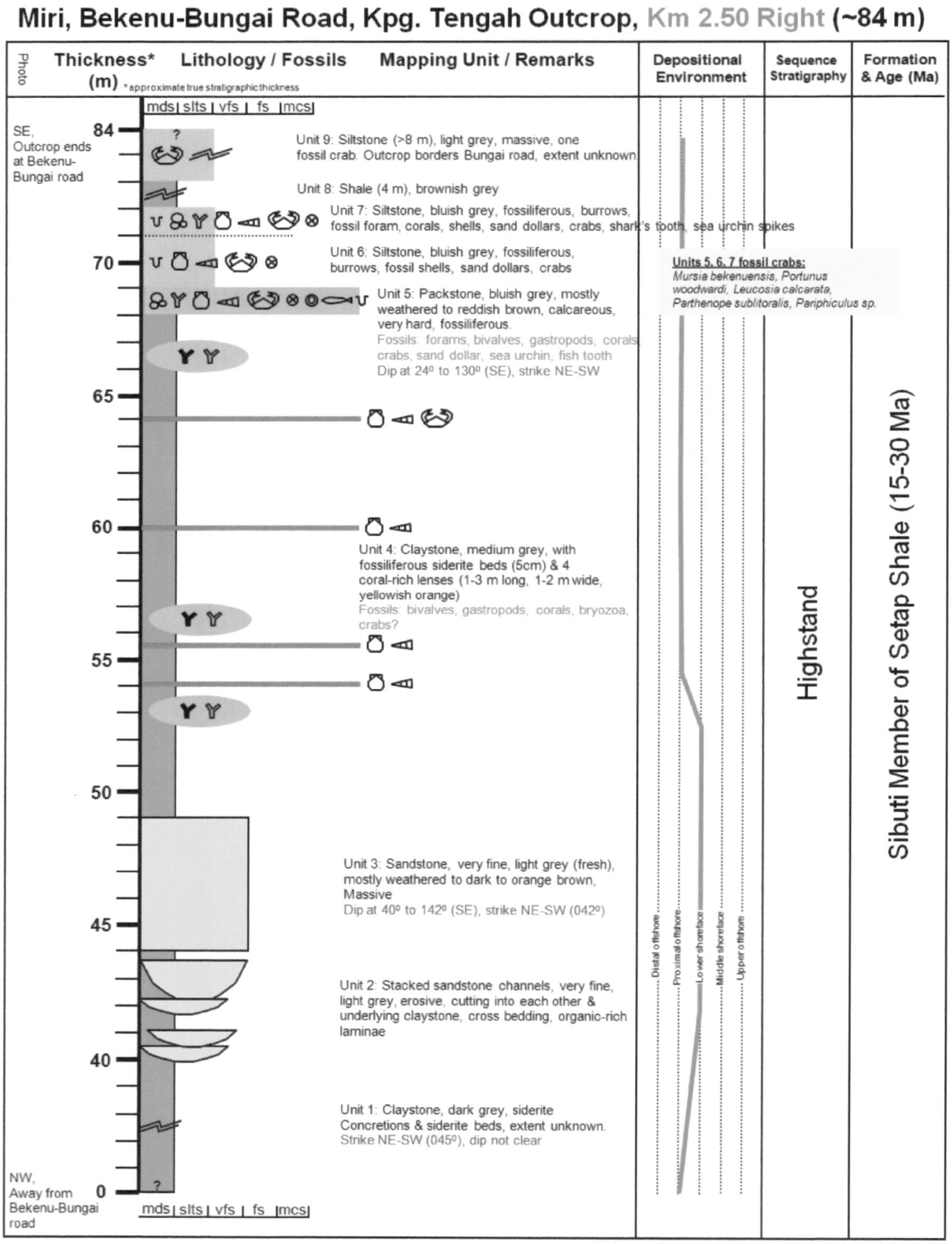

Miri, Bekenu-Bungai Road, Kpg. Tengah Outcrop, Km 2.50 Right **(~84 m)**

Figure 4.7.4a: Stratigraphic log of the Kampong Tengah outcrop

4.7.4 Geological Description

The Bungai Road branches off from the main coastal road and continues toward the Bungai Beach on the west. The road cuts through the gentle coastal plain, consisting of siltstones and mudstones of the Sibuti Formation (Figure 4.3a). Fossil-rich beds are exposed by road cuts and excavations at four locations (Figure 4.7.1a), the main outcrop being located at km 2.5R (also called the Kampong Tengah outcrop).

The Kampong Tengah outcrop, now an abandoned and partly excavated hill, exposed a thick sequence of calcareous siltstones, bioclastic packstones, mudstones, and sandstones of the Sibuti Formation. The fossils are concentrated in a 4-meters thick interval, comprising three distinct beds made up of siltstone and bioclastic packstones (Figure 4.7.4a). These beds have yielded a rich assemblage of fossils including crabs, corals, sea urchins, gastropods, bivalves, fish bones and teeth, and shark's teeth (Plates 4.7.4b,c). Fossil crabs *Mursia bekenuensis*, named after the nearby town of Bekenu, were first discovered here in 1996.

Further along the road there are small outcrops on both sides (km 1.4). These have yielded rich mollusk fauna (Plate 4.7.4c, Figures 1-6), review of which is in progress. The description of the remarkable gastropod *Melongena murifactor* (Plate 4.7.4c, Figures 3-4) is based on collections from these outcrops.

Plate 4.7.4b: Fossils from the Sibuti Formation

Crabs: Figure 1: *Mursia bekenuensis* (Collins, Lee & Noad, 2003), Bungai road km 2.50 right (Kampong Tengah), a-b dorsal view, pincer; Figure 2: *Charybdis annulata* (Fabricius, 1798), Sibuti-Niah road km 6.4 left, a-b, dorsal and ventral views; Figure 3: *Raninoides morrisi* (Collins, Lee & Noad, 2003), Sibuti-Niah road km 6.3 right, dorsal view; Figure 4: *Parthenope sublitoralis* (Morris & Collins, 1991), Bungai road km 2.50 right (Kampong Tengah), dorsal view; Figure 5: *Portunus woodwardi* (Morris & Collins, 1991), Bungai road km 2.50 right (Kampong Tengah), dorsal view; Figure 6: *Pariphiculus sp.*, Bungai road km 2.50 right (Kampong Tengah), dorsal view; Figure 7: *Leucosia calcarata* (Collins, Lee & Noad, 2003), Bungai road km 2.50 right (Kampong Tengah, dorsal view; Figure 8: *Carcinoplax prisca* (Imaizumi, 1961), Bungai road km 2.50 right (Kampong Tengah), dorsal view;

Other fossils: Figure 9: Sand dollar, Bungai road km 2.50 right (Kampong Tengah); Figure 10: Heart-shaped sea urchin, Sibuti-Niah road km 6.3 right; Figure 11: Round sea urchin, Bungai road km 2.50 right (Kampong Tengah); Figure 12: Shark's teeth, Bungai road km 2.50 right (Kampong Tengah).

Plate 4.7.4c: Fossils from the Sibuti Formation, Kampong Tengah

Gastropods. Figure 1: *Xenophora (Xenophora) solaroides* (Reeve, 1845) – these animals agglutinate other shells and materials onto their shell for camouflage – km. 1.4; Figure 2: *Haustellum mindanaoensis* Dohrn, 1862 – km 1.4; Figures 3-4: *Melongena murifactor* (Vermeij & Raven, 2009) – which has an apertural septum as can be seen in the view from above - Holotype and paratype – km 1.4; Figure 5: *Pugilina spec.* – km 1.4; Figure 6: *Oliva (Galeola) mitrata* (Martin, 1879) – km 1.4 Sea urchin. Figure 7: 'sand dollar' type – km 2.5;

Solitary corals: Figure 8: *Heterocyathus sp.?*, a parasitic worm which lived in the small opening; Figure 9: *Dendrophyllia sp.?*; Figure 10: *Flabellum sp.?*; Figure 11: *Trachyphyllia / Catalaphyllia ?*; Figure 12: *Cycloseris ? (Fungiidae).*

4.8 Sibuti-Batu Niah Coastal Road

4.8.1 Access and Location

From the Miri town centre (Clock Tower roundabout, km 0.0), the drive to Sungai Sibuti bridge (sign posted) is about 45 km (Figure 4.3a). The start of the fossil-rich exposures is about 5 km SW of the bridge, on both sides of the road (km 5.0 to km 7.9). Individual fossil outcrops are indicated in Figure 4.8.1a.

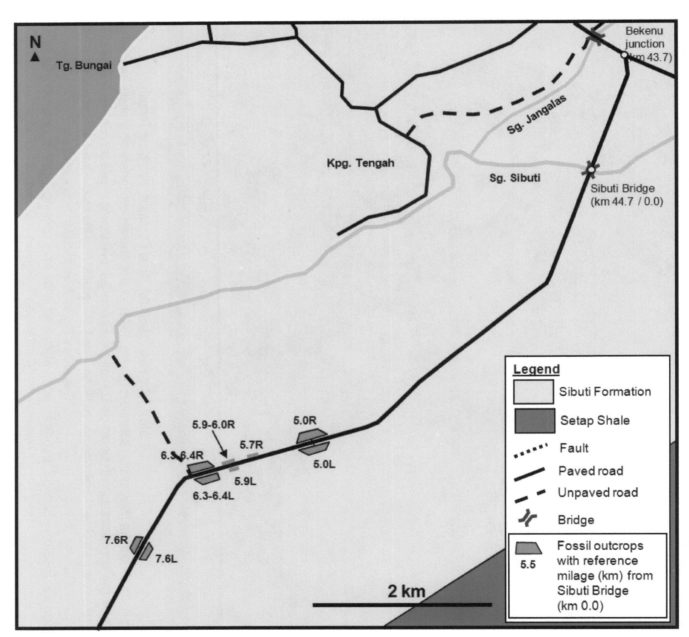

Figure 4.8.1a: Location and geological map of fossil localities, Sibuti-Batu Niah Road

4.8.2 Logistics and Safety

Road safety is a particular concern at this location. The coastal road is narrow, almost always with soft and steep edges, and very limited space/location to park safely. To get to the outcrops some walking along the side of the road is necessary, so watch out for passing vehicles (often at high speed!), especially when crossing the road.

4.8.3 Outcrop Highlights

This an easily accessible fossil site, with fossil-rich road cuts along the coastal road. Here you will find an abundance of fossil crabs, with the crabs *Charybdis annulata* being most common, followed by *Raninoides morrisi* and *Mursia bekenuensis*. Other fossils include shark's teeth, sand dollars, sea urchin, gastropods and bivalves.

4.8.4 Geological Description

The Sibuti-Batu Niah coastal road cuts through gentle, rolling hills, in the area southwest of Bekenu town and Sungai Sibuti Bridge (Figure 4.3a). Along this stretch of the coastal road, the rocks mainly consists of mudstones and siltstones of the Sibuti Formation (Figure 2.2a), which grades into the Setap Shale further southwest. The main fossil outcrops, exposed in road cuts, are found between km 5.0 to km 7.9. The best fossil outcrops are at km 6.4L&R, km 6.3R and km 5.0L&R.

Fossils found at this location are similar to those at km 2.5R (Kampong Tengah outcrop) on Bungai road and at km 38.9R on Beraya-Bekenu road. In particular, fossil crabs *Mursia bekenuensis* and *Raninoides morrisi*, and sand dollar fossils can be found at all three outcrops (Plate 4.7.4b, Figure 9).

No detailed geological map is available for this location, and therefore the distribution of fossils in the outcrops has not been worked out. Limited dip measurements indicate that the coastal road may have cut through a series of small, southwest plunging anticlines.

4.9 Brunei Tektites: Glass Stones from the Sky

Although tektites are not organic fossil remains, they are treated in this chapter because they are valuable collectables of a major meteorite impact.

About 700,000 to 750,000 years ago, a large meteorite crashed onto the Earth, somewhere in Southeast Asia; the impact crater has not been identified, probably because it lies on the seabed. The evidence for this hypervelocity impact comes from melted and quenched terrestrial rocks, called tektites, that were blown away or sucked into the vacuum created by the meteorite and were dispersed over an extended geographical area, called a strewn field. The "Australasian strewn field" covers an area of some 50 million km2, which makes it the largest strewn field on Earth, extending from South China to Australia.

Tektites are not evenly dispersed over a strewn field, but have a showery distribution pattern. The largest specimens, sometimes weighing more than a few kilograms, are found in the Northern hemisphere, while smaller ones are generally found at the far southern end of the strewn field; these must have entered the uppermost atmosphere, as they show signs of re-melting as they reentered the lower atmosphere. As a result of the extreme pressures and temperatures experienced by these rocks, all constituent minerals form a homogeneous melt of glass from which all water was eliminated; tektites range amongst the driest rocks on Earth! Their color is black but in thin section they are brown and their shining surfaces are pitted and/or grooved due to acid etching in the soil. They are round, lozenge- or pear-shaped and often flattened.

Figure 4.9a: Largest tektite specimen from Butir, about 5.5 cm long and 178 g in weight (Collection Jan Wilschut)

Figure 4.9b: Largest tektite specimen from Tutong, 4.3 cm long and 68 g in weight (Collection Jan Wilschut)

Swiss geologist F.P. Mueller first reported tektites from Brunei in 1913 during his fieldwork for the British Borneo Petroleum Syndicate Ltd. He found four specimens in white sands on a hill beyond the bazaar in Tutong, and sent them to Basle for an analysis of their chemical and physical properties. The results were published in 1915 and remain a landmark report in tektite studies.

It was not until 1961 that G.E. Wilford, a geologist working for the Geological Survey Department of the British Territories in Borneo reported the presence of tektites at gravel workings in the Barakas and Butir areas. At Pentuan Hill he collected a specimen in-situ, which he sent for potassium-argon (K-Ar) dating by J. Zahringer in Vienna who derived an age of about 730,000 years.

Following a visit to Brunei, Prof. G.H.R. von Koenigswald (1961) compared the Brunei tektites from Butir with similar tektites found in Java (Indonesia), where they occur in the Trinil beds near Sangiran together with an abundance of fossils of early-middle Pleistocene age.

In 1970, Brunei State Geologist R. B. Tate described a collection of tektites from Butir, collected by Mr. Hee Kui Fong, including a specimen weighing 300 grams, being the largest recorded tektite found in Brunei to date. This specimen compares well with the Bikol specimen found in the Coco Grove, Paracale area in the Philippines, and with other Philippine rizalites from the Stamesa area.

In 1975 T. Harrison compared age results from shells, charcoal and wood from NW Borneo (Niah and Sabah) with similar materials found in tektite-bearing terraces at Butir in Brunei. In 1978 Brunei Shell geologist J.G. Wilschut described and illustrated tektites from his collections from Butir, Jerudong and Tutong (Figures 4.9a,b,c) and showed a picture of the beach outcrop at Jerudong where tektites can be collected (Figure 4.9d). Most tektite specimens are only a few millimeters in size, and centimeter-sized specimens are rare and prized by collectors (Figure 4.9e).

Figure 4.9c: Tektite with grooved surface ("schlieren" patterns), Butir (Collection Jan Wilschut)

Tektites in northern Borneo fell as a shower only in an area of Brunei, roughly from Berakas to Tutong, during or shortly after the deposition of the Jerudong terrace. Sediments consist of pebble beds and white leached beach sands, which stretch all along the coast from Muara to Bintulu in Sarawak. Erosion of the uplifted Jerudong terrace, especially in the Jerudong area, resulted in the presence of reworked tektites in much younger terraces in the Butir and Pentuan area. Erosion continues to the present day, especially in the Jerudong area, and to a lesser extent near Tutong, where tektites can be found on the beaches. Tektites can also be found in old gravel workings in the Butir area.

Figure 4.9d: Tektite in situ, within pebble bed; Jerudong Beach (Photograph by Jan Wilschut)

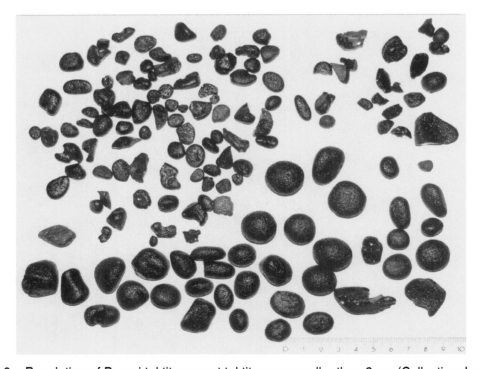

Figure 4.9e: Population of Brunei tektites; most tektites are smaller than 2 cm (Collection Jan Wilschut)

Appendix to Chapter 4: Checklist of Fossil Crabs

Some 31 species of fossil crabs from 24 genera have been recorded from the area around Miri. Two of these crabs are named after their discovery locations: *Phylira trusanensis* (after Tanjung Tusan, named "Trusan" in older maps) and *Mursia bekenuensis* (after Bekenu town).

The taxonomic classification of the identified fossil crabs is given below.

Kingdom: Animalia

Phylum: Arthropoda

Superclass: Crustacea

Order: Decapoda

Infraorder: Brachyura

Family: *LEUCOSIIDAE* Samouelle, 1819
    Genus: *NUCIA* Dana, 1852
        Species: *Nucia platyspinosa* Collins, Lee & Noad, 2003
        Species: *Nucia borneoensis* Morris & Collins, 1991
    Genus: *PHILYRA* Leach, 1817
        Species: *Philyra trusanensis* Collins, Lee & Noad, 2003
        Species: *Philyra granulose* Morris & Collins, 1991
    Genus: *PARIPHICULUS* Alcock, 1896
        Species: *Pariphiculus papillosus* Morris & Collins, 1991
        Species: *Pariphiculus multituberculatus* Collins, Lee & Noad, 2003
    Genus: *IPHICULUS* Adams & White, 1848
        Species: *Iphiculus sexspinosus* Morris & Collins, 1991
    Genus: *MYRA* Leach, 1817
        Species: *Myra subcarinata* Morris & Collins, 1991
    Genus: *DRACHIELLA* Guinot in Serene & Soh
        Species: *Drachiella guinotae* Morris & Collins, 1991
    Genus: *LEUCOSIA* Weber, 1795
        Species: *Leucosia calcarata* Collins, Lee & Noad, 2003
    Genus: *NUCILOBUS* Morris & Collins, 1991
        Species: *Nucilobus symmetricus* Morris & Collins, 1991

Family: *PORTUNIDAE* Rafinesque-Schmaltz, 1815
    Genus: *PORTUNUS* Weber, 1795
        Species: *Portunus woodwardi* Morris & Collins, 1991
        Species: *Portunus obvallatus* Morris & Collins, 1991
    Genus: *CHARYBDIS* de Haan, 1833
        Species: *Charybdis annulata* (Fabricius, 1798)
    Genus: *PODOPHTHALMUS* Lamarck, 1801
        Species: *Podophthalmus vigil* (Fabricius, 1798)

Genus: *SARATUNUS* Collins, Lee & Noad, 2003
        Species: *Saratunus longiorbis* Collins, Lee & Noad, 2003
Genus: *SCYLLA*
        Species: *Scylla serrata* (Forsskal, 1755)

Family: *PILUMNIDAE* Samouelle, 1819
Genus: *GALENA* de Haan, 1833
        Species: *Galena obscura* (H. Milne Edwards, 1834)
        Species: *Galena litoralis* Collins, Lee & Noad, 2003

Family: *CALAPPIDAE* de Haan, 1837
Genus: *CALAPPA* Weber
        Species: *Calappa sexaspinosa* Morris & Collins, 1991
Genus: *MURSIA* Leach in Desmarest, 1823
        Species: *Mursia bekenuensis* Collins, Lee & Noad, 2003

Family: *GONEPLACIDAE* MacLeay, 1838
Genus: *CARCINOPLAX* H. Milne Edwards, 1852
        Species: *Carcinoplax prisca* Imaizumi, 1961

Family: *HEXAPODIDAE* Miers, 1886
Genus: *HEXAPUS* Miers, 1886
        Species: *Hexapus decapodus* Morris & Collins, 1991

Family: *MACROPTHALMIDAE* Dana, 1851
Genus: *MACROPHTHALMUS* Latreille in Desmarest, 1823
        Species: *Macrophthalmus (Mareotis) wilfordi* Morris & Collins, 1991

Family: *RANINIDAE* de Haan, 1839
Genus: *RANINOIDES* H. Milne Edwards, 1837
        Species: *Raninoides morrisi* Collins, Lee & Noad, 2003

Family: *DORIPPIDAE* MacLeay, 1838
Genus: *HEIKEA* Holthuis & Manning, 1990
        Species: *Heikea tuberculata* Morris & Collins, 1991

Family: *MAJIDAE* Samouelle, 1819
Genus: *PISOIDES* H. Milne Edwards & Lucas, 1803
        Species: *Pisoides cf. bidentatus* (H. Milne Edwards, 1873)

Family: *PARTHENOPIDAE* MacLeay, 1838

Chapter 5

Preserving the Geological Heritage

5.1 Towards Protection of Geological Sites in Miri

Miri is renowned for the first commercial oil field in Malaysia, discovered in 1910. It is also home to interesting geological exposures. This chapter summarizes efforts made by a group of volunteers to preserve a selected number of these exposures as permanent geological exhibits that also support educational and eco-/geo-tourism efforts.

The Airport Road Outcrop in Miri, located about two kilometers from the city center along the route to Miri Airport, is one of the most outstanding geological exposures in the area. Petroleum companies operating in the region routinely make geological field trips to this site. In addition, the site has attracted university teachers and students from Malaysia and overseas. Its location on one of the highest points in Miri also offers visitors a panoramic view of the city.

In this chapter we review the work done by the Miri Outcrop Museum Working Group since its inception in 1998. The Miri Outcrop Museum Working Group is a volunteer group that came together to preserve some of the outcrops recognized to be excellent geological sites. Since 1998, the composition of the Working Group has changed with some members leaving and new ones joining. The goals and spirit of the group has stayed firmly on track.

5.2 Objectives of the Project

Like in any other project, a plan was set up and a series of prioritized actions established. Distilling the initial ideas and developing them into a practical concept formed the first step. It was realized from the outset that a project like this could not be sustained simply by a group of volunteers but would need an ongoing maintenance effort that could only come from local authorities. Hence, it was a very clear requirement to involve relevant authorities as early as possible.

It was therefore essential to present results of the effort to local dignitaries and the Miri Council very early on in the project. The support and response to this work has been extremely positive. It was also recognized that concentrating on geological aspects alone would not be at all sufficient. Initial ideas were further developed to integrate the geological goals into the aspirations of the community in terms of eco-tourism and education.

The main objectives of the Miri Outcrop Museum Project are:
- Preserve key geological outcrops in and around Miri
- Make these easily accessible and provide appropriate information
- Encourage and conduct educational tours for schools, tourists, and the public
- Work with appropriate bodies (the Miri Municipal Council and the Geological Survey) for sustainability

- Link with tour operators to provide added eco-touristic attractions for Miri
- Link with institutes of higher learning

5.3 Selection Criteria for the Airport Road Outcrop

The Miri Airport Road Outcrop is a significant geological location for several reasons. Along a series of cut surfaces (Figure 5.3a), it provides opportunities to study sedimentological and structural geological features in 3 dimensions. In particular, the outcrop has been hailed to be of exceptional quality to study fault development in plan and vertical view over distances often exceeding 50 meters (a suitable reservoir scale). Lateral variations along fault planes and presence of remobilized clay sealing fault compartments are of special importance for fundamental research in rock fracturing. For the oil industry, clay smear along fault planes and deformation bands on the sides of faults are especially important to establish the sealing capacity of faults and permeability barriers within the sandstones. The outcrop provides excellent analogies for faulted sandstone reservoirs in oil fields, in particular the Miri oil field and offshore Sarawak. Its location on one of the highest points in Miri offers visitors a panoramic view of the city and the western coastline.

The geology of the Airport Road Outcrop is described above in Chapter 3.4. Over the years, this outcrop has been the focus of multiple studies (Burhannudinnur & Morley, 1997; Sorkhabi & Hasegawa, 2005), including a PhD project (van der Zee, 2001). A geological guide on a CD entitled "*Geology of the Miri Airport Road outcrop*" has been developed by Lesslar & Wannier (2001) and is available at the website http://www.ecomedia-software.com

Figure 5.3a: Overview of the Miri Airport Road outcrop

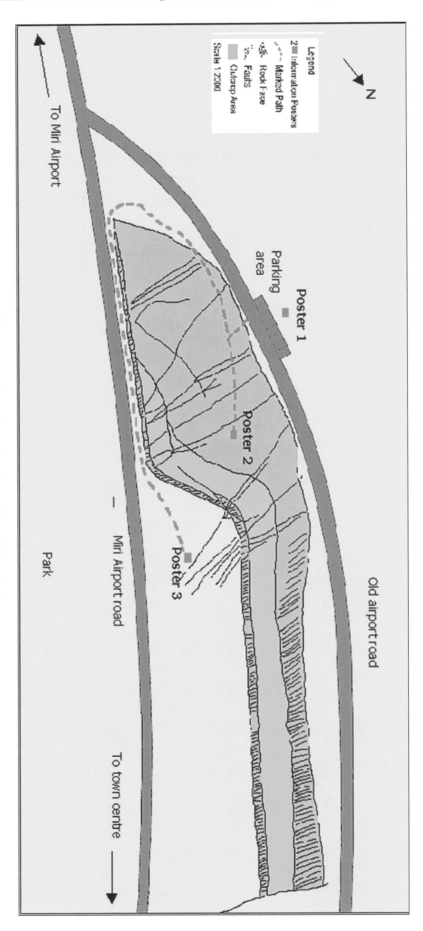

Figure 5.4a: Location of posters, Airport Road outcrop

5.4 Implementation

Following the approvals and financial commitment by the Miri Municipal Council, a number of key things were implemented at the outcrop. Guided excursions were planned, and it was the intention that anyone visiting the outcrop on their own should be able to follow a well-marked trail and read the explanations for important geologic features on a set of posters. Three geological posters were prepared and mounted permanently at the site. Apart from information about the outcrop, these posters also gave contact information should visitors to the site wish to pursue the topic further. The locations of these posters at the outcrop are shown in Figure 5.4a. Poster 1 is located at the top of the outcrop, next to a parking area, where an open view of the coastline can be gained (Figure 5.4b); Poster 2 is located centrally on the outcrop (Figure 5.4c); and Poster 3 is located at the base of the outcrop, facing its most interesting feature, the faulted face of the outcrop (Figure 5.4d).

The Miri Municipal Council had also been agreeable to the construction of a concrete footpath as well as handrails (Figure 5.4e) for increased safety in the sloping sections. The footpath greatly increases the accessibility of the base of the outcrop which previously had only been accessible through some sometimes long grassy and muddy pathways; it makes traversing the outcrop a much more pleasant experience.

Figure 5.4b: Poster 1

Figure 5.4c: Poster 2, Airport Road outcrop

Figure 5.4d: Poster 3, Airport Road outcrop

Figure 5.4e: Handrail to facilitate the visit of Airport Road outcrop

Figure 5.4f: Flyers for schools, Airport Road outcrop

Since the outcrop museum became functional, the working group has focused mainly on field excursions for secondary school students in order to increase the local students' awareness of geology and the environment. In October 2000, the first of these school group excursions was conducted. Some 20 students from SMK St Joseph Miri were given the inaugural tour of the outcrop and their reactions and questions were noted down.

Pictures and details of this and other excursions are presented on a website (http://www.ecomedia-software.com/museum/museum.htm). This website offers information on the history of this project as well as on the progress of activities. In addition, an information flyer was produced (Figure 5.4f) and distributed to schools, the Miri Municipal Council and Tourist Information Centers.

5.5 Ongoing Challenges

Over time, weather and vegetation takes its toll on geological exposures, especially in the tropical climate of Malaysia. Sections of the outcrop have already crumbled and other parts are overgrown with vegetation. General maintenance and safety of the outcrop and the surrounding areas are not always ideal. An ongoing supervision is required and this is not optimal with a volunteer group, whose members not only have full time occupations but also change every so often. Finding a proper home for the supervision and maintenance of the outcrop has therefore been a high priority item for the volunteer group.

The CD entitled *Geology of the Miri Airport Road Outcrop* was our attempt to preserve the many interesting facets of the outcrop in pictures and detailed descriptions. While the many excursions already conducted have proved to be valuable, a lot more remains to be done to preserve the geological and historical Miri Airport Road Outcrop. This effort serves as a good example and learning set for the study and preservation of similar outcrops in and around Miri for geologists, the public and visitors to the area.

The twelve years that have elapsed since the inception of this idea in 1998 have been very interesting and rich in learning. At the beginning, there were more questions than answers, but through organizing and leading excursions, many of the issues were resolved. It has been very rewarding to see excitement on the faces of school children, most of whom have not had exposure to geology before, suddenly realizing that they are standing on something very important in their own city. It has been very satisfying to receive the collaboration of colleagues and the backing of our ministers and the Miri Municipal Council, whose support has enabled this project to come to fruition.

5.6 Sustaining the Effort

The completion of the posters, stands and footpath were essential milestones in getting the site ready for visitors. Some publicity including communications with schools was the next step in our effort. The preservation of these geological sites has limited value if it is only for the sake of preservation; education and ecotourism are equally important priorities. The following ongoing activities were therefore done to help towards these aims:

(1) Creating awareness among the local residents especially schools and travel agents;

(2) Working with the Miri Council on handling of issues such as waste disposal, flyer distribution, additional construction requirements;

(3) Working with other bodies such as the geological survey and local universities regarding geological aspects, and

(4) Developing an information package that can be used independently by travel agents.

Over the past 10 years since the first tour was conducted, several more have been organized for local schools. These were led by members of the volunteer group and were well received. However, the geological knowledge of the group was not easy to maintain as members transferred and changed out from time to time. Hence efforts were made from the outset to find a proper governing body for this work.

On 23rd April, 2010, a meeting was held between the Outcrop Museum Working Group and the Minerals and Geosciences Department of Malaysia, Sarawak GeoHeritage section (JMG). The meeting was also attended by a representative of the Miri City Council and supported by the presence of Yang Berhormat (YB) Andy Chia, State Legislative Council Member for Pujut. During this meeting, the JMG agreed to take over the technical management of the outcrop, with the Miri City Council providing assistance for the maintenance of the outcrop site. As the office of the JMG is in Kuching, the Miri-based volunteer group will continue to provide help in conducting tours as and when possible.

5.7 Acknowledgements

The Miri Outcrop Museum Project gratefully acknowledges the following persons for their enthusiasm and encouragement: (1) YB Datuk Patinggi Tan Sri Dr George Chan Hong Nam, Deputy Chief Minister, Minister of Tourism, and Minister for Industrial Developments, Sarawak (2) YB Mr Lee Kim Shin, Member of State Legislative N65 Senadin and Assistant Minister for Infrastructure Developments, and Social Developments Sarawak; (3) Dato' Wee Han Wen, Former Mayor of Miri; (4) Professor H.D.Tjia, of the LESTARI Group; (5) Staff of Miri Municipal Council; and (6) many volunteers who helped with the project in various capacities.

Chapter 6

Glossary of Geological Terms

This glossary gives simple definitions for some of the geologic jargon for non-geologists. It is admittedly incomplete, and interested readers should consult a geology dictionary.

Accretionary prism: a thick wedge of often faulted sediments, forming at a subduction zone.

Aggradation (aggrading): vertical stacking of similar beds or of beds deposited within the same environment.

Alluvium: recent sediments deposited on land, mainly referring to unconsolidated fluvial and alluvial sediments.

Anticline (anticlinal): an elongated domal structure formed by tectonic compression and bending of sedimentary layers and/or basement rocks.

Antithetic faults: a set of faults dipping against a major fault at a considerable angle; opposite of synthetic faults.

Bedforms: sedimentary bodies reflecting the flow conditions at the time of deposition, such as ripple marks, sand waves and dunes.

Benthic or benthonic: organisms living on sea floor.

Bioclasts (bioclastic): sedimentary particles derived from the breakdown of sea shells.

Biofacies: A body of sediments in the stratigraphic record characterized by the assembly of particular fossils and their ecology.

Biostratigraphy: an aspect of stratigraphy dealing with the relative dating of sedimentary rocks, using marker fossils (usually microfossils).

Bioturbation: displacement of sedimentary particles within the seabed as a result of burying and moving organisms.

Biozone (biozonation): a relative geological time unit characterized by the occurrence of a marker species or a combination of marker species, generally microfossils.

Bottomsets: a series of the horizontal or gently dipping beds deposited in front of a deltaic basin.

Boulder: rounded clastic particles larger than 25cm (the largest clastic component of river bed sediments).

Burrow (burrowing): tunnel or hole made within sediments by an animal; the fossilized traces of a moving animal within the sediment.

Calcite (calcitic, calcareous): carbonate mineral formed of calcium carbonate/calcite (CaCO3).

Carbonaceous: sediments rich in carbon, usually coaly particles.

Carbonate: a rock made chiefly of carbonate minerals such as calcite (CaCO3).

Cenozoic: the Era of Earth history from the end of the Cretaceous (65.5Ma) to present day. It encompasses the Tertiary and the Quaternary periods.

Chronostratigraphy: stratigraphy dealing with the relative or numerical (absolute) dating of sedimentary rocks (using the methods of biostratigraphy, tephrachronology, isotope stratigraphy, and radiometric geochronology).

Clastics (clasts): sedimentary rocks composed of particles derived from weathered rocks

Clay (claystone): hydrated, plastic and less permeable sediment consisting of clay minerals; claystone is hardened and compacted clay.

Coarsening-upward cycle: a suite of sedimentary beds, typically sandstones, characterized by a vertical increase in their grain-size. Coarsening upward cycles are generally interpreted as progressively shallowing-upward of the sedimentary environment.

Column: as used in cave geology, the merged pair of stalactite and stalagmite, forming a continuous column from cave floor to ceiling.

Compressional deformation: deformation resulting from tectonic stress generated from the convergence of two tectonic plates. Reverse faults, thrust faults, and folds are common compressional features.

Conglomerate (conglomeratic): sedimentary rock composed of large rounded particles (in the range of granules to boulders) embedded in finer sediments.

Continental block: an individual tectonic block with a granitic basement.

Continental collision: tectonic collision between two continental blocks.

Continental crust: generally the granitic basement of continents.

Cretaceous: period of Earth history from the end of the Jurassic (145.5 Ma) to the beginning of the Cenozoic (65.5 Ma).

Crossbedding (crossbeds): a sedimentary structure in which the beds lie oblique to each other. High-energy depositional environments with shifting currents often result in sandstone made up entirely of sets of oblique sand laminae, frequently scouring and cutting each other.

Crust: the outermost solid layer of the Earth, resting on the mantle; the crust can be continental or oceanic.

Damage-zone (fault damage-zone): a zone of rock along a fault plane or between two closely-spaced fault planes that has been fractured and deformed by shear.

Deformation-band: thin (mm-wide) deformation plates in sandstone fault-damage zones. Deformation bands form in highly porous sandstones: grain-to-grain sliding, rotation and grain fracture during fault shearing result in compacted and cemented sand plates with little or virtually no porosity or permeability.

Delta (deltaic): wedge-shaped sedimentary landform at the mouth of a river (where the rivers flows into a lake, sea or an ocean).

Diagenesis (diagenetic): secondary (post-depositional) chemical and physical (overburden compaction) processes that take place in sediments and ultimately convert the soft sediments into a sedimentary rock. Major chemical diagenetic processes include cementation, dissolution, re-crystallization and replacement.

Diapir (diapiric): vertical rise and intrusion of a remobilized mass of low-density, under compacted sediments (notably salt and clay) that cut through and bend the overlying denser sediments. In Sarawak and Brunei, the Setap shales have been found in the cores of such diapirs.

Dike: a tabular body of rock that penetrates the overlying strata along discrete vertical pathways (a dike may be an igneous rock or vertically injected sediments).

Dinoflagellates (cysts): group of marine and lacustrine unicellular organisms that create an organic membrane (cyst). Remnants of dinoflagellate cysts within fine-grained sediments can be used for dating the sediments.

Dip: magnitude of the inclination of a bed, measured in degrees by a geological compass.

Dipmeter: An instrument to measure along-hole dips during well drilling.

Distal: far away; opposite of proximal.

Distributary channels: in a river delta, streams that branch off from the main river course.

Drag (fault drag): Bending of strata at the contact with a fault. The orientation of the drag indicates the direction of fault movement.

Draperies: flowstone mineral formation on cave ceilings and walls, characterized by thin, curtain-like, wavy sheets of calcite hanging downward.

Drift: in tectonics, the movement of a tectonic plate relative to others.

Dripstones: a general term for cave deposits formed by mineral crystallization from water dripping from cave ceiling.

Ductile: characteristic of soft rock formations, such as claystone and salt, which can deform by flow without fracturing; opposite of "brittle".

Embayment: a body of shallow water indenting the shoreline towards the land. Marine embayments are generally protected from coastal currents but are strongly affected by tides. Brunei Bay is an example of an estuarine embayment.

Eocene: an epoch of the Paleogene (Early Tertiary) from the end of the Paleocene (55.8 Ma) to the base of the Oligocene (33.9 Ma).

Extensional deformation: deformation resulting from tectonic processes whereby two blocks move away from each other, usually at an angle. Open fractures, and normal faults forming graben and horsts are common extensional features.

Facies: a set of descriptive-interpretational attributes such as lithological (lithofacies) or biological (biofacies) that characterizes sedimentary rocks in terms of their depositional and biological environments.

Fault (faulting): a sharp break in the continuity of rock bodies caused by tectonic stress release. A fault plane marks the displacement of two rock columns which have moved relative to each other. "Normal faults" are extensional structures, which may be steep (high angle) or very low angle (detachment normal faults). "Syn thetic" normal faults slip parallel to each other. "Antithetic" normal faults slip against each other. "Listric faults" (spoon-shaped structures) are normal faults which flatten at the base. "Growth faults" are extensional faults forming during sediment deposition (therefore, sediment thicknesses on the downthrown sides are thicker than the corresponding beds on their upthrow sides). "Reverse" and "thrust" faults are compressional structures; reverse faults are high angle, but thrust faults are low angle (30° or less). "Strike-slip" or "transform" faults form when blocks move laterally (past each other).

Fining-upward: a sedimentary cycle characterized by vertical decrease in grain-size. Fining-upward cycles are generally interpreted as progressively deepening-upward of the sedimentary environment.

Flame structure: a soft-sediment deformation shown as an A-shaped structure in the bedding.

Flaser-bedding: a sedimentary pattern consisting of alternating sand and mud layers, formed by intermittent flows in tidal environments. Flasers correspond to sandy ripples, draped by mud layers when water currents abate.

Flooding surface: a bed boundary marking a sudden deepening in depositional environment.

Flowstones: cave deposits consisting of calcitic sheets, mounds and curtains, formed by thin water films flowing along cave walls.

Footwall block: in tectonics, a fault block beneath the fault plane; opposite of "hanging wall block."

Foraminifera: a group of marine, microscopic, unicellular organisms in which the cell is protected by tiny shell made of calcite or agglutinated sediment particles. Foraminifera are adapted to either a planktonic or a benthonic mode of life. Some larger foraminifera can reach cm-sizes. Planktonic foraminifera are used for strati graphic dating while benthonic foraminifera also give information about the depo sitional environment and water depth.

Foresets: a series of inclined beds deposited on the slope of a delta or inclined laminae deposited along the slope of a bedform.

Foreshore: high-energy zone of sedimentation on the shore that is influenced by the rise or fall of tides.

Fossils: remains or imprints of dead animals and plants preserved in sediments.

Gouge (fault gauge): clay-rich, fine grained, crushed rock produced by friction along a fault plane when the two fault blocks move relative to each other.

Graben: a faulted depression bounded by two normal faults. Opposite of horst.

Granules: rounded clastic particles larger than sand and smaller than pebbles; of a size between 2 and 4mm.

Gravel: clastic sedimentary rock composed dominantly of granules and pebbles.

Hanging block: in tectonics, a fault block overlying the fault plane; opposite of "footwall block."

Helictites: a form of stalactites characterized by erratic growth, including sideward and U-turn vertical growth.

Herringbone cross bedding: a sedimentary structure consisting of alternating, opposing cross beds. Reversing currents in tidal environments commonly cause bipolar orientation of foresets.

Heterolithics: lithofacies characterized by a mixture of sandstone, siltstone and shale that is commonly bioturbated.

Highstand (sea level): high phase in a cycle of (relative) sea-level variations. During highstands, the continental shelves are generally flooded.

Holocene: the youngest epoch in Earth history, starting 11,700 years ago and continuing to present. The Holocene is an inter-glacial period that followed the Weichselian glaciation.

Horsetail structures: in tectonics, horse-tail like splaying of fractures at the tip of a fault.

Horst: An upthrown land block bounded by two normal faults. Opposite of graben.

Hummocky cross bedding (hummocks): sedimentary structures characterized by cross-cutting "smile"-like shapes, developed in sandstones and formed by the action of large storms.

Imbrication (sedimentology): a sedimentary structure found in very coarse-grained river sediments and characterized by pebbles inclined in the same direction (shingle structure). The oblique orientation of the pebbles is due to strong river currents; imbrications point to the orientation of the currents: pebbles dip towards the current and face upward towards the sea.

Inversion tectonics: reactivation of a fault in a style opposite to its former deformation (for example, a normal fault reactivated as a reverse fault, or vice versa).

Joints: A set of extensional fractures dissecting a rock, showing no vertical displacement.

Karst (karstification): limestone country affected by surface and subsurface dissolution processes resulting in the development of a major subsurface aquifer. Caves, sinkholes, and underground rivers characterize a karst country. Mulu and Niah caves in Sarawak were formed by karstification.

Langhian: lower stage of the Middle Miocene between 16 Ma and 13.8 Ma.

Lenticular bedding: a sedimentary pattern consisting of isolated lenses of sandstones in a mudstone matrix. Lenticular bedding characterizes relatively deep or distal intertidal environments.

Limestone: chemical or biological sedimentary rock made up essentially of calcite mineral ($CaCO_3$).

Limonite (limonitic): yellowish-brown to black iron oxide minerals formed by weathering (oxidation) of iron-bearing minerals. It occurs as coatings or earthly masses in many sandstone formations around Miri.

Lithofacies: A mappable sedimentary facies in the stratigraphic record characterized by lithology.

Marl: calcium carbonate or lime-rich mudstone.

Maximum flooding surface: a surface recording the maximum paleo-bathymetry within a unit of sediments.

Mesozoic: the Era of Earth history encompassing the Triassic, Jurassic and Cretaceous periods (251 – 65.5 Ma).

Messinian: uppermost stage of the Miocene from 7.2 Ma to 5.3Ma.

Meta-sedimentary rock: sedimentary rock that has undergone some form of metamorphism at low temperatures (examples include phyllite, quartzite, and marble).

Micropaleontology: a branch of paleontology (in contrast to macropaleontology) dealing with the study of fossil microorganisms.

Miocene: an epoch of the Neogene (Late Tertiary) spanning from the end of the Oligocene (23.03 Ma) to the base of the Pliocene (5.3Ma). Most of sedimentary rocks in Sarawak are Miocene in age.

Mud (mudstone): a fine-grained sedimentary rock consisting of a mixture of clay and silt-sized particles. Hardened and fissile (platy) mudstones are described as shales.

Nannofossils / Nannoplankton (calcareous nannoplankton): marine, mainly tropical unicellular microscopic organisms with shells composed of tiny, calcium carbonate plates. The study of the crystal structure and shape of these plates help to assess the biostratigraphic age and depositional environment of sediments.

Neogene: middle period of the Cenozoic stretching from the end of the Paleogene (23.03 Ma) to the beginning of the Quaternary (2.6 Ma).

Oceanic crust: generally the basaltic crust of the ocean basins.

Overbank sediments: fine-grained sediments, clay, silt, and fine sands deposited when rivers overflow their banks.

Packstone: a type of limestone consisting dominantly of shell particles cemented by fine mud.

Paleontology: a branch of geology dealing with the study of fossils and life evolution through Earth's history.

Palynomorphs: a general term for microscopic particles of organic material such as spores and pollen, found in sediments.

Paralic: low-energy coastal environments such as swamps and lagoons.

Pebbles: rounded clastic particles larger than granules and smaller than cobbles, of a size between 4 and 64 mm.

Pellets (peloids): mm-scale oval particle, possibly the excreta of organisms such as crabs.

Phanerozoic: ("the appeared life") the last 542 million years of Earth history, from the beginning of the Cambrian to present time, abundant in larger life forms.

Phyllite: foliated coarse-grained low-temperature metamorphic rock.

Plankton (Planktic or planktonic): organisms living in suspension in the water column and moving passively, depending on currents.

Pleistocene: the main epoch of the Quaternary, from 2.65 Ma to 0.01 Ma, during which Earth experienced the latest glacial-interglacial cycles.

Pliocene: the youngest epoch of the Neogene, ranging from 5.3 to 2.6 Ma.

Pop-up structure: in tectonics, a relief created when a block is compressed laterally and uplifted along two reverse faults.

Progradation (prograding): vertical stacking pattern of beds showing a shift to progressively shallower environment, such as the outbuilding of a delta.

Quaternary: the most recent period of Earth history, starting 2.65 Ma and comprising the Pleistocene and the Holocene.

Resurgence: karstic spring (2)

Sandstone: a detrital (clastic) sedimentary rock composed essentially of cemented sand grains.

Schist: foliated fine-grained high-temperature metamorphic rock.

Sediments (sedimentary rocks): rock-forming materials deposited on the surface of the continents, river beds, lake or oceans and formed by weathering and erosion of older rocks, or by chemical-biological precipitation in water bodies.

Sedimentology: a branch of geology that studies the composition and formation of sediments and processes of sedimentation in a basin.

Seep (oil seepage): naturally-occurring outflow of hydrocarbon to Earth surface.

Sequence boundary: a surface separating two major stratigraphic sequences, each characterized by a different history of sedimentation.

Serravallian: upper stage of the Middle Miocene between 13.8 Ma and 11.6 Ma.

Shale: Hard and fissile (platy) mudstone

Shoreface: steeply sloping zone between the seaward limit of the shore at low water and the nearly flat inner shelf zone.

Siltstone: a sedimentary rock with a grain size in the silt range (between 1/16 and 1/256 mm), finer than sandstone and coarser than claystone (mudstone and shale).

Sinkhole: a natural depression in the surface topography caused by removal by water and collapse of bedrock, and connecting the surface to underground caves.

Speleothems: a general term for secondary mineral cave deposits, such as stalactites, stalagmites, and flowstones.

Stalactites: elongated calcitic mineral formations hanging from cave ceilings. Stalactites are often aligned with joints or fractures as these structures facilitate water percolations.

Stalagmites: elongated calcitic mineral formations protruding vertically from cave floor. Stalagmites are often formed below and in line with stalactites when there is excess water running down from the ceiling.

Stratification: layering of sedimentary rocks, whereby individual beds are separated by mappable planes, and reflecting changes in the composition and depositional conditions of sediments.

Stratigraphy: a branch of geology dealing with the arrangement and age succession of sedimentary rocks.

Strike: a line representing the intersection of a bedding plane with a horizontal plane, and measured as a deviation from the North Pole.

Strike-slip fault: fault characterized by displacement of rocks along a horizontal plane; the movement may be right-lateral (sinistral) or left-lateral (dextral).

Subduction (subduction zone): a process at convergent plate boundaries where an oceanic tectonic plate descends beneath a continental plate.

Subsidence: downward movement of Earth surface; opposite of uplift.

Suture, suture zone: in tectonics, a zone along which two continental plates have collided.

Syncline: downward-curving fold, with rock layers dipping toward the center of the structure. A syncline is generally developed between two adjacent anticlines.

Syn-sedimentary: refers to processes happening as sediments are being deposited.

Tectonics: a branch of geology dealing with the description of deformations and structures affecting rocks, and the origin of forces and movements that have created these structures. Tectonics usually deals with structures on a regional scale or long-term geologic development, while structural geology investigates individual structures in details or related structures in a given area.

Tertiary: The largest period of the Cenozoic, extending from the end of the Cretaceous (65.5Ma) to the beginning of the Pleistocene (2.6Ma). The use of the terms Tertiary and Quaternary, although well known in geologic literature, is discouraged by geological circles because these terms are remnants of a classification of Earth history into four periods, now outdated.

Throw (fault throw): the amount of vertical displacement of rocks by a fault.

Thrust fault: fault characterized by the upward movement of one rock block with respect to another block. Thrust faults are low-angle reverse faults; both reverse and thrust faults create a positive relief (uplift).

Tidal flat: a transitional environment between a fully continental and a fully marine environment, and found in many estuarine areas. Tidal flats are typically covered by high tide waters and exposed during low tides.

Tortonian: lower stage of the Upper Miocene between 11.6 Ma and 7.2 Ma.

Transgression (transgressive): land-ward shifting of shoreline as the sea spreads laterally; a marine transgression on the geologic record occurs when sea-level rises relative to the land.

Transpression (transpressional fault): a fault having components of strike-slip and shortening (compression). Such a fault may form regionally by oblique subduction or collision of plates, or locally as a restraining bend during the side-way stepping of a strike-slip fault. On seismic sections, transpression gives rise to a positive flower structure.

Transtension (transtensional fault): a fault having components of strike-slip and extension which occurs either regionally by oblique normal faults in rift zones or locally as releasing bends during the side-way stepping of a strike-slip fault. On seismic sections, transtension gives rise to a negative flower structure.

Trough cross bedding: bedding structure characterized by curved laminae that fill in elongate scours.

Unconformity: erosional surface separating two sets of strata of different ages. Unconformities are said to be conformable when the two sets of strata are parallel to each another; in angular unconformities, however, one set of strata lies in a considerable angle to the other sets of strata above and below it.

Uplift: vertical elevation of Earth surface in response to natural forces, mainly tectonic faulting and folding.

Upthrown block: a rock block that has experienced an oblique vertical translation along a fault plane.

Weichselian: the most recent glacial period occurring at the end of the Pleistocene, spanning approximately between 110,000 and 11,700 years ago.

Chapter 7

References & Bibliography

APA Publications, 1988. Insight Pocket Guides: Sarawak. APA Publications, Singapore, 95 p. plus map.

Artis, L.C., 1941. The Paleontology of the Miri Field. Shell company report, issued 31 Jan. 1941.

Beets, C., 1947. Note on fossil Echinoidea and Gastropoda from Sarawak and Kutei, Borneo. Geologie & Mijnbouw, 9: 40-42.

Burhannudinnur, M., and Morley, C.K., 1997. Anatomy of growth fault zones in poorly lithified sandstones and shales: implications for reservoir studies and seismic interpretation: part 1, outcrop study. Petroleum Geoscience, 3: 211-224.

Caline, B., and Huong, J., 1992. New insights into the recent evolution of the Baram Delta from satellite imagery. Geological Society of Malaysia Bulletin, 32: 1-13.

Collins, J.S.H., Lee, C., and Noad, J., 2003. Miocene and Pleistocene Crabs (Crustacea, Decapoda) from Sabah and Sarawak. Journal of Systematic Palaeontology, 1: 187-226.

Farrant, A.R., Smart, P.L., Whitaker, F.F., and Tarling, D.H., 1995. Long-term Quaternary uplift rates inferred from limestone caves in Sarawak, Malaysia. Geology, 23: 357-360.

Geocon (B) Sdn. Bhd., 1992. Kajian Perpindahan Kauri Tunggulian (Sand). Technical Report for the Public Works Department, Ministry of Development, Brunei Darussalam.

Gow, D., 1941. Economic Prospects of the Miri Field. Shell company report, issued February 1941.

Haile, N.S., 1962. The Geology and Mineral Resources of the Suai-Baram Area, North Sarawak. Memoir 13, Geological Survey Department British Territories in Borneo, Kuching, 176 p.

Haile, N.S., 1969. Geosynclinal theory and the organizational pattern of the north-west Borneo geosyncline. Quarterly Journal of Geological Society, London, 12: 171-194.

Haile, N.S., 1974. Borneo. In: Spencer, A.M. (ed.) Mesozoic-Cenozoic Orogenic Belts. Geological Society, London, Special Publication 4, 333-347.

Halim, Q.A., 1996. Development of Non-metallic Mineral Resources in Brunei. Ministry of Development, Public Works Department, Brunei Darussalam.

Hall, R., 1996. Reconstructing Cenozoic SE Asia. In: Hall, R. and Blundell, D. (eds.) Tectonic Evolution of Southeast Asia. Geological Society, London, Special Publication 106, 153-184.

Harper, G.C., 1975. The Discovery and Development of the Seria Oilfield. Muzium Brunei, Brunei Darussalam, 99 p.

Harrison, T., 1975. Tektites as "date Markers". Asian Perspectives, 18(I): 61-63.

Hazebroek, H.P., and Kashin bin Abang Morshidi, A., 2000. National Parks of Sarawak. Natural History Publications (Borneo). Kota Kinabalu, 502 p.

Ho, K.F., 1978. Stratigraphic framework for oil exploration in Sarawak. Geological Society of Malaysia Bulletin, 10: 1-13.

Hose, C., 1927. Fifty Years of Romance and Research, or A jungle-wallah at Large. Hutchison & Co. Ltd, London, 301 p.

Hose, C., 1929. The Field-book of a Jungle-wallah. Witherby, London, 216 p.

Howell, G., 1926. Oil development, storage, transport and distribution. British Empire Oilfields Sarawak and Trinidad. Oil Engineering and Finance, May 1926, 209-212.

Hutchison, C.S., 1989. Geological Evolution of South-East Asia. Oxford University Press, 368 p.

Hutchison, C.S., 2005. Geology of North West Borneo: Sarawak, Brunei and Sabah. Elsevier, Amsterdam, 421 p.

Ibrahim, A., 1998, Depositional History of the Liang Formation in Brunei Darus salam and Implications for Petroleum Geology. Research Report of the Petroleum Geoscience Faculty at Universiti Brunei Darussalam.

Jackson, J.C., 1968. Sarawak: A geographic survey of a developing state. University of London Press, 218 p.

Koenigswald, G.H.R. von, 1961. Tektites in Borneo and elsewhere. Sarawak Museum Journal, vol. 10 (no. 17-18, July-December) (new series), 319-324.

Lalanne de Haut, J.P.Y.M., van Delden, J.M., and van Niel, J.P., 1968. Recent Sedimentation in the Baram Delta and Adjacent Areas. Brunei Shell Petroleum Company Limited/ Sarawak Shell Limited, Exploration Department, Geological Report 892.

Lesslar, P., and Wannier, M., 1998. Destination Miri – A geological tour Northern Sarawak's National Parks and Giant Caves. Ecomedia Software CD-rom. http://www.ecomedia-software.com

Lesslar, P., and Wannier, M., 2001. Geology of the Miri Airport Road Outcrop. Ecomedia Software CD-rom. http://www.ecomedia-software.com

Lesslar, P. and Lee, C. 2001. Preserving our key geological exposures - exploring the realm of geotourism. Warisan Geologi Malaysia (Geological Heritage of Malaysia), 4: 417-426.

Liechti, P., Roe, F.N., Haile, N.S., and Kirk, H.J.C., 1960. The Geology of Sarawak, Brunei and the Western Part of North Borneo. Bulletin 3, Volumes 1&2. Geological Survey Department British Territories in Borneo. Government Printing Office, Kuching, Sarawak, 360 p.

Madon, M. B.Hj., 1999b. Geological Setting of Sarawak. In: The Petroleum Geology and Resources of Malaysia. PETRONAS, Kuala Lumpur, 273-290.

Martin, K., 1925. Bericht over fossielen van Sarawak. Reports for the Bataafsche Petroleum Maatschappij No. 18 (manuscript dated 20 April 1925).

Martin, K., 1926. Bericht over fossielen van Tutong (Brunei) en Miri (Sarawak). Reports for the Bataafsche Petroleum Maatschappij No. 20 (manuscript dated 4 April 1926).

Martin, K., 1931. Bericht over Tertiaire verstuwingen van Brunei. Reports for the Bataafsche Petroleum Maatschappij No. 21 (manuscript dated 4 December

1931).

Martin, K., 1932. Tweede bericht over fossielen van Brunei, Reports for the Bataafsche Petroleum Maatschappij No. 22 (manuscript dated 5 February 1932).

Meredith, M., Woodbridge, J., and Lyon, B., 1992. Giant Caves of Borneo. Tropical Press Sdn. Bhd., Kuala Lumpur, Malaysia, 142 p.

Morley, C.K., Back, S., van Rensbergen, P., Crevello, P., and Lambiase, J.J., 2003. Characteristics of repeated, detached, Miocene-Pliocene tectonic inversion events, in a large delta province on an active margin, Brunei Darussalam, Borneo. Journal of Structural Geology, 25: 1147-1169.

Morris, S.F., and Collins, J.S.H., 1991. Neogene crabs from Brunei, Sabah and Sarawak. Bulletin of the British Museum Natural History (Geology), 47(1): 1-33.

Mueller, F.P., 1915. Tektites from British Borneo. Geological Magazine, 2: 206-211.

Nuttall, C.P., 1961. Gastropoda from the Miri and Seria Formations, Tutong Road, Brunei. In: Wilford, G.E (ed.) The Geology and Mineral Resources of Brunei and Adjacent Parts of Sarawak. Memoir 10, British Borneo Geological Survey, 73-87.

Payne, R., 1960. The White Rajahs of Sarawak. Funk & Wagnall, New York, 274 p.

PETRONAS, 1999. The Petroleum Geology and Resources of Malaysia. PETRONAS, Kuala Lumpur, 665 p.

Pybus, C., 1997. The White Rajahs of Sarawak: Dynamic Intrigue and the Forgotten Canadian Heir. Double & McIntyre, Vancouver, 235 p.

Raven, J.G.M., 2002. Notes on molluscs from NW Borneo. 1. Stromboidea (Gastropoda, Strombidae, Rostellariidae, Seraphidae). Vita Malacologica, 1: 2-32.

Raven, J.G.M., 2008. Fossil Shells. In: McIlroy, R., and N.E. Yusniasita Dols (eds.) The Seashore Life of the Brunei Heart of Borneo, Vol. 3. The Seashells. Panaga Natural History Society, Seria, 60-63.

Reinhardt, M.,Wenk, E., 1951. Geology of the colony of North Borneo. Bulletin of the Geological Survey Department of the British Territories in Borneo 1, 160 p.

Reece, R.H.W., 1982. The Name of Brooke: The End of White Rajah Rule in Sarawak. Oxford University Press, Kuala Lumpur. 331 p.

Rijks, E.J.H., 1981. Baram Delta geology and hydrocarbon occurrence. Geological Society of Malaysia Bulletin, 14: 1-18.

Runciman, S., 1960. The White Rajahs: A History of Sarawak from 1841 to 1946. Cambridge University Press, Cambridge, 319 p.

Sandal, S.T., 1996. The Geology and Hydrocarbon Resources of Negara Brunei Darussalam (1996 revision). Brunei Shell Petroleum Co. and Muzium Negara, Syabas Bandar Seri Begawan, Brunei Darussalam, 243 p.

Sawata, H., 1991. Note on fossil locality of crabs and cuttlefish bones in Pleistocene marine sandy clay at Sungai Kolok, southeast-most Thailand. Warta Geologi, 17(1): 1-10.

Sarawak Shell Berhad, Public Affairs Department, 1990. The Miri Story: The Founding Years of the Malaysian Oil Industry in Sarawak, 3rd ed. Sarawak Shell Berhad,

Lutong, Sarawak, 74 p.

Schumacher, P. von, 1941. The Geology and Prospects of the Miri Field. Shell company report, issued 7 February 1941.

Schweitzer, C.E., Scott-Smith, P.R., Ng, and P.K.L., 2002. New occurrences of fossil decapods crustaceans (Thalassinidae, Brachyura) from late Pleistocene deposits of Guam, United States Territory. Bulletin of Mizunami Fossil Musuem, 29: 25-49.

Shuib, M.K., 2001. The Miri Structure – a dextral strike-slip model. Petroleum Geology Conference & Exhibition 2001, Paper 20, Kuala Lumpur, Malaysia, 84-87.

Simmons, M.D., Bidgood, M.D., Brenac, P., Crevello, P.D., Lambiase, J.J., and Morley, C.K., 1999. Microfossil assemblages as proxies for precise paleoenvironmental determination – an example from Miocene sediments of northwest Borneo. In: Jones, R.W., and Simmons, M.D. (eds.) Biostratigraphy in Production and Development Geology. Geological Society, London, Special Publication 152, 219-241.

Sorkhabi, R., 2010. Miri 1910: The centenary of the Miri discovery in Sarawak. Geo Expro, March 2010, 44-49.

Sorkhabi, R., and Hasegawa S., 2005. Fault Zone Architecture and Permeability Distribution in the Neogene Clastics of Northern Sarawak (Miri Airport Road Outcrop), Malaysia. In: Sorkhabi, R., and Tsuji, Y. (eds.) Faults, Fluid Flow, and Petroleum Traps. AAPG Memoir 85: 139-151.

Tate, R. B., 1970. Tektites in Brunei. The Brunei Museum Journal, 2(1); 253-263.

Tingay, M.R.P., Hillis, R.R., Swarbrick, R.E., Morley, C.K., and Damit, A.R., 2009. Origin of overpressure and pore-pressure prediction in the Baram Province, Brunei. AAPG Bulletin, 93: 51-74.

Van der Zee, W., 2001. Dynamics of Fault Gouge Development in Layered Sand-Clay Sequences. Lehr- und Forschungsgebiet fur Geologie – Endogene Dynamik der RWTH Aachen, 155 p.

Vermeij G.J. & J.G.M. Raven, 2009. Southeast Asia as the birthplace of unusual traits: the Melongenidae (Gastropoda) of northwest Borneo. Contributions to Zoology, 78: 113-127.

Wannier, M., 2009. Carbonate platforms in wedge-top basins: An example from the Gunung Mulu National Park, Northern Sarawak (Malaysia). Marine and Petroleum Geology, 26: 177-207.

Wilford, G.E., 1961. The geology and mineral resources of Brunei and adjacent parts of Sarawak with description of Seria and Miri Oilfields. Memoir Geological Survey Department 10, British Territories in Borneo, Brunei 10, 319 p.

Winkler Prins, C.F., 1996. Dr C. Beets (1916-1995) and the 'Rijksmuseum van Geologie en Mineralogie'. Scripta Geological, Leiden, 113: 1-21.

CPSIA information can be obtained
at www.ICGtesting.com
Printed in the USA
LVHW011649260623
750800LV00002B/25